SCIENTIFIC AND EDUCATIONAL JOURNAL OF THERAPEUTIC RIDING

2015

ISSN 1174-6238

ISBN 978-0-9845490-9-2

A PUBLICATION OF

THE FEDERATION OF HORSES IN EDUCATION AND THERAPY INTERNATIONAL A.I.S.B.L.

The Federation of Horses in Education and Therapy International A.I.S.B.L

SECRETARIAT
HETI
P.O. Box 134, Damariscotta, ME 04543
Email: office@HETIfederation.org
www.HETIfederation.org
www.facebook.com/HETI.international

ISSN 1174-6238
ISBN 978-0-9845490-9-2

Table of Contents

Dear HETI Members

We are thrilled to be able to tell you that you are holding the 20th issue of the Scientific and Educational Journal of Therapeutic Riding in your hands. This is a proud moment in the history of this publication and in the life of The Federation of Horses in Education and Therapy International.

And what a journey it has been since the Journal's debut in 1994, containing the proceedings of the VIII Congress in New Zealand. We've leafed through some of the back issues the other day and found a wide variety of subjects and approaches:

- Articles that sought to capture the effects of different models like Therapeutic Riding, Hippotherapy, Remedial/Adaptive/Psycho-motor Therapeutic Vaulting, and Equine-Facilitated Psychotherapy, and
- Articles that described sessions using one of these models with many different physical or mental challenges like Cerebral Palsy, MS, ADD and ADHD, Autism, Blindness, Breast Cancer Survivor, Spinal cord injury, Head injury, Learning Disabilities, Tetraplegia etc.
- Horses: in earlier journals general description of the characteristics of horses and their senses dominated, In later journals we found research regarding stress on horses during therapy sessions and the effects of saddle fit and rider weight on the natural movement of the horse
- Research: from information on how to write an article for publication and how to write a research proposal, to articles about research methods and the state of research in hippotherapy, plus many, many case studies
- Finally, we found contributions that spoke from the perspective of the rider, such as Para-equestrian in competition

From humble beginnings this journal has evolved into a true peer-reviewed scientific organ on the forefront of scientific research in equine therapy. This has been due in no small part to the tireless efforts of everyone involved.

We would like to thank the authors, the peer reviewers, the copy editors and especially the editors. What a wonderful, dedicated group of people:

Richard Jaroszewiscz (Belgium), 1994, 1996

Dr. Gillian Lawrence (Canada), 1994, 1996, 1997

Mary Longden (Australia), 1994, 1996, 1997, 1998, 1999, 2000

Jean Tebay (USA), 1994, 1996, 1997, 1998, 1999, 2000

Dr. Olive Webb (New Zealand), 1994, 1996, 1997, 1998, 1999, 2000, 2001, 2002, 2003

Dr. Carl Klüwer (Germany), 1998, 1999, 2000

Isabelle Quinet (Belgium), 1998, 1999, 2000

Dr. Beth L. Macauley (USA), 2005, 2006, 2007

Sanna Mattila-Rautiainen (Finland), 2008, 2009

Ann Kern-Godal (Norway), 2010, 2011

Nina Ekholm-Fry (USA), 2012/13, 2014, 2015

Sue Cotton (Australia), 2012/13, 2014, 2015

The XV Congress this past June in Taiwan was an affirmation of the unity and enthusiasm of our members. We look forward with great anticipation to your participation in the XVI Congress in Ireland (Dublin) in 2018 and invite you to visit the XVI Congress website at www.heti2018.org for further information.

Please read this Journal, use it as a resource, share it with your colleagues, and reference it often. Our best wishes and appreciation go to each one of you for your continued support.

Sanna Mattila-Rautiainen
President, HETI (The Federation of Horses in Education and Therapy International A.I.S.B.L.)

Gisela Rhodes
Executive Director, HETI (The Federation of Horses in Education and Therapy International A.I.S.B.L.)

Editorial

Welcome to the 20th issue of the Scientific and Educational Journal of Therapeutic Riding! The SEJThR is a scholarly, international journal that specifically focuses on horses and other equines in therapy, rehabilitation, and education/learning settings.

One of the highlights of the research year 2015 was the 10-year anniversary of the Horses and Human Research Foundation. In the past decade, this foundation, dedicated to funding research projects that explore the benefits of horses and their potential to impact on the health and wellness of people, awarded close to half a million U.S. dollars to research projects, which were selected from applicants from 18 countries worldwide. Foundations like HHRF are central to the development and sustainability of our field as they assist with a crucial element for research: funding.

Another research-related highlight this year was the largest randomized control trial so far in our field, conducted by Gabriels and colleagues. Randomized Controlled Trial of Therapeutic Horseback Riding in Children and Adolescents with Autism Spectrum Disorder was published in the Journal of the American Academy of Child & Adolescent Psychiatry. It included several elements necessary for a good-quality scientific study: a large sample (116 participants), randomized assignment to intervention or control, and test raters blinded to participant assignment. In addition, the program utilized for the study was a longer program (10 weeks).

For us at SEJThR, 2015 meant continuing the success of the 2014 journal updates, among those a double-blind peer-review system, which supports the necessary rigor of a scholarly journal. We welcome more individuals interested in joining our international body of reviewers! Please visit our Call for Reviewers on page 9 or go to the journal's website at www.hetifederation.org/journal for more information and how to support our field in this valuable way.

For the year ahead we wish to focus on two journal areas that directly serve practitioners around the world: the case study and our theory and practice section. A case study can serve to describe the reasoning behind, and application of, a treatment or educational strategy involving equines, as well as its outcomes on an individual level. The theory and practice section includes submissions ranging from discussions of relevant topics to proposals for research protocols.

Original research, such as pilot studies, and systematic literature reviews are always welcome. Please review our submission guidelines on page 8 or go to www.hetifederation.org/journal for more information. We welcome first-time authors and are available for questions at editorial@hetifederation.org.

The SEJThR creates an excellent opportunity for international information sharing, and it is our hope that it serves as an important platform for future studies and connections around the world.

In this issue:

- **Johnson et al.** presents two case studies on the effects of therapeutic riding on two individuals on the autism spectrum
- **Lac** explores challenges to the daily lives of families affected by autism through a case study
- **Tonteri** discusses the potential of the horse in adolescence aggression management
- **Doner and Ekholm Fry** discusses the relevance of learning theory in the handling and training of the therapy horse
- **Barnfield** responds to Anestis et al. (2014) in a letter to the editor.

As always, please direct your emails with comments and questions to editorial@hetifederation.org.

Sincerely,

Nina Ekholm Fry (U.S.A./Finland) and Sue Maree Cotton (Australia)

Call for Manuscripts

The Federation of Horses in Education and Therapy International (HETI) publishes the Scientific and Educational Journal of Therapeutic Riding (SEJThR) annually. We are currently seeking submissions with the purpose of advancing knowledge in the fields of therapeutic riding and equine-assisted activities and therapies.

The SEJThR has several submission categories:

Original Research
An original research study conducted by the author

Review
A comprehensive or systematic review of research on a specific topic

Case Study
A presentation of a single comprehensive case study or multiple case studies

Theory and Practice
A discussion of a specific topic in the field

Letter to the Editor
Comments on articles published in the SEJThR

Manuscripts are reviewed through a double-blind peer-review system and submissions are welcome throughout the year.

For more information about manuscript requirements and the submission process, please visit www.hetifederation.org/journal.html.

Please contact journal editors Nina Ekholm Fry and Sue Maree Cotton with questions.

Journal Editors:
Susan Maree Cotton • Nina Ekholm Fry
editorial@hetifederation.org
www.hetifederation.org

Call for Reviewers

The Federation of Horses in Education and Therapy International (HETI) publishes the Scientific and Educational Journal of Therapeutic Riding (SEJThR) annually. The focus of this journal is on advancing knowledge in the fields of therapeutic riding and equine-assisted activities and therapies.

Reviewers are needed to ensure the quality and status of the SEJThR as a peer-reviewed journal. Peer-reviewers conduct blind reviews of manuscript submissions within their area of knowledge as a step in the review process.

Expectations for reviewers are:

- Ability to provide constructive feedback for a scholarly manuscript in terms of its quality and contribution to the field
- Timely and respectful communication with journal editors

Requirements for reviewers are:

- Professional experience in an area of therapeutic riding/equine-assisted activities and therapies
- A master's or doctoral degree
- Experience in reviewing scholarly texts
- Ability to complete a review within four weeks of receiving a manuscript

If you are interested in contributing your time as a peer-reviewer for the SEJThR, please submit a letter of interest and your curriculum vitae (CV) per email to editorial@hetifederation.org. Please type "Reviewer Application" in the subject line.

Letter to the Editor

The Effectiveness of Equine-related Therapies: A Response to Anestis, Anestis, Zawilinski, Hopkins and Lilienfeld (2014)

Anne M. C. Barnfield , Brescia University College

Correspondence to:
Dr. Anne Barnfield, Associate Professor, Psychology,
Brescia University College, 1285 Western Road, London, Ontario, Canada, N6G 1H2.
Email: abarnfie@uwo.ca
Phone: (519) 432-8353, ext. 28246

Keywords

Equine-related therapies, Research, Methodology, Psychological benefits.

THE EFFECTIVENESS OF EQUINE-RELATED THERAPIES: A RESPONSE TO ANESTIS, ANESTIS, ZAWILINSKI, HOPKINS AND LILIENFELD (2014)

The recent article by Anestis, Anestis, Zawilinski, Hopkins and Lilienfeld (2014) in the Journal of Clinical Psychology has caused a reaction in many who are involved in therapies using equines and in the related research fields. In their review article, Anestis et al. (2014) assess a number of research studies on mental health therapies involving equine partners, describing and evaluating each study and giving their judgment regarding the validity of the studies. In the overall conclusion of the article it is stated that "current evidence does not justify the marketing and utilization" (Anestis et al., 2015, p. 1115) of equine-related therapies.

I was alerted to this article by the December newsletter of Horses in Education and Therapy International (HETI, formerly FRDI), where the article's conclusion was noted with consternation (Kern Godal, 2014). The Horses and Humans Research Foundation (HHRF) newsletter of December 2014 also noted the conclusions of this study, with a call for more research in this field, not less (HHRF, 2014a). I, and many others, would agree that rather than ceasing what Anestis et al. (2014) call "equine-related treatments (ERT)" (p. 1115), there is a need to expand research in this area and for that research to be published.

Reading the article by Anestis et al. (2014) left me with the impression that the authors had taken a particular position - in this case that equine related treatments are ineffective - then examined available studies looking for errors and potential problems with the research. Even if this were not the case, in my opinion Anestis et al (2014) are not being reasonable: they expect perfection, requiring the "gold standard" in a very difficult area for research methodology.

There is a great deal of anecdotal and research evidence in favour of equine-related therapies and reports of beneficial effects are found in media ranging from newspapers and popular magazines (e.g., McVeigh, 2012; Wilkins, 2014) to the peer review journals which published the articles evaluated by Anestis et al. (2014) (e.g., Journal of Autism and Developmental Disorders; Psychiatric Rehabilitation Journal). Scientific research into equine related therapies is still in its infancy however and, yes, there is room for improvement, but that is not to say that such therapy is to be avoided. After all, how will research into effectiveness be run if the therapies are abandoned?

THE EFFECTIVENESS OF EQUINE-RELATED THERAPIES: A RESPONSE TO ANESTIS, ANESTIS, ZAWILINSKI, HOPKINS AND LILIENFELD (2014)

I am perhaps biased, working in this field myself, but I believe that many of the issues noted by Anestis et al. (2014) are simply inherent to the area of study. Although Anestis and colleagues do acknowledge difficulties of research in equine-related therapies, they still seem to require impossible standards. For example, although it is the ideal, often one cannot have a treatment control group because of difficulties of assembling a matched group of such individuals or ethical issues regarding withholding of treatment[1]. My own work is on the effects of Therapeutic Riding (TR) for children (e.g., Carey, Murray & Barnfield, 2013). This type of research uses a form of convenience sampling where we have to study those who are available to us and cannot assign participants to categories at our convenience. In this research, for a control group we would have had to age and condition-match all children who had signed up to participate in TR sessions. At the time there was no waitlist to use as controls, and the children's physical and mental conditions varied. In such a situation, it thus verges on the impossible to create a valid control group. We could ask to hold back half of the children who have registered for TR, but no sensible parent is going to sign consent for such a study. Who would deny their child what is at the very least an enjoyable activity, and potentially much more, for the convenience of an unknown researcher? The most appropriate option, then, is to use each child as their own control. Baseline, pre-riding session behaviours are measured and then compared to post-riding session scores to make within-subjects comparisons. Not perfect methodology, perhaps, but the best that can be done in this situation.

Similarly, whilst having observers/research assistants blind to the study hypothesis might be the gold standard for observational research, in practice this is often impossible to achieve. Unless one has a large grant and numerous assistants available, having a person who does not know the hypothesis under investigation perform observations really is not an option. Research is often performed by those who have devised the study simply because there is no-one else to do so. It could be argued (as do Anestis and colleagues) that the researcher is potentially biased by their understanding of the study's hypothesis and potential outcomes, but it is possible to set up a study utilizing methods known to reduce possible biases in recording of results. Research can be kept as objective as possible by, for example, use of rigourous checklists and behavioural definitions which have been finalized well before any observations are made. Making such

[1] Some research does allow for a treatment control group, when participants are first recruited from a specific population and then assigned treatment or not (e.g., Gabriels, Pan, Dechant, Agnew, Brim, & Mesibov, 2015), but many established therapeutic situations – as with the TR study described here – cannot support this.

observations pre-and post-treatment with no reference to the original record sheets also reduces potential for bias.

One benefit of ERT which is often overlooked, and in fact listed as a failing by Anestis et al. (2014), is that such activities tend to be viewed positively by participants. People who may not feel motivated to participate in other forms of therapy can be encouraged to participate when horses are involved (e.g., Beetz, Winkler, Julius, Uvnäs-Moberg, & Kotrschal, 2015). As noted by the HHRF recently: "Motivational aspects of equine therapy are an assumed strong impact factor in the therapy session" (2014b, n.p.). Children particularly may not even see activities such as TR as being therapy. A parent interviewed for the qualitative aspect of our TR research stated that the best part of her daughter's TR participation was the child was happy to attend and did not perceive the session as therapy. Another parent pointed up the "normalcy" aspect, noting that: "The riding gives her an activity she can participate in that levels the playing field with other kids." Taking part in activity that other youngsters might also do gives children something to talk about with peers, thus incidentally decreasing social isolation. These motivational aspects of ERT can be viewed as advantages, rather than methodological failings. Another "failing" which could actually be viewed as an advantage is that participants might be interested in the opportunity to interact with the equine partners. This "novelty effect" was listed as another methodological issue by Anestis and colleagues, rather than being seen as a potential benefit. Surely something which encourages participation in therapy should be viewed as a positive influence rather than a negative? How one could eradicate any novelty effects of therapy with equines is also an interesting question - one suggestion from a colleague (given in jest) was: "Well, you could try using cows!" In seriousness, some studies have used comparisons of interactions with equines versus being in a similar atmosphere (e.g., barn work; Gabriels et al, 2015) or comparing use of dogs to horses (e.g., Nurenberg et al., 2015), but to say that in all situations one could eliminate or control for the novelty of ERTs is, again, to require the impossible.

Anestis et al. (2014) do allow that there are "stubborn obstacles" (p. 1129) and difficult methodological issues in equine-related therapy research, but then to assume that all such research is therefore invalid is too absolute. There is much good which comes from equine-related activities in therapeutic settings and to dismiss outright the utility of any such intervention is an extreme position. I have neither the time nor the expertise to run the sort of extensive analysis that Anestis and colleagues have done - perhaps others working in this field could do a similar study to look, as suggested by Kern-Godal (2014), for the good achieved by such work, rather than only the flaws in its execution. More funding will need to be provided to run such trials. It is unfortunate, however, that there is reluctance on the part of funding bodies to give grants for such research. At least here in Canada there seems to be a form of negative, circular reasoning shown in response to grant applications: No funding for research until we know this type of therapy is legitimate; until research shows this to be legitimate, no funding.

Until research into equine-related treatments is more fully funded it will be difficult to eliminate sources of bias. Even then, with issues such as some of the examples given above, research studies can never be perfect; we will simply have to do the best we can. I for one strongly support the need for further studies in this field, and will be continuing my own research into the psychological effects of therapeutic riding.

References

Anestis, M. D., Anestis, J. C., Zawilinski, L. L., Hopkins, T. A., & Lilienfeld, S. O. (2014). Equine-related treatments for mental disorders lack empirical support: A systematic review of empirical investigations. Journal of Clinical Psychology, 70 (12), 1115-1132.

Beetz, A., Winkler, N., Julius, H., Uvnäs-Moberg, K., & Kotrschal, K. (2015). A comparison of equine-assisted intervention and conventional play-based early intervention for mother-child dyads with insecure attachment. Journal of Occupational Therapy, School, and Early Intervention, 8, 17-39.

Carey, A. D., Murray, S. J. & Barnfield, A. M. C. (2013). The effects of therapeutic riding on children. Scientific and Educational Journal of Therapeutic Riding, 2012/13, 10-35.

Gabriels, R. L., Agnew, J. A., Holt, K. D., Shoffner, A., Zhaoxing, P., Ruzzano, S., Clayton, G. H., & Mesibov, G. (2015). Pilot study measuring the effects of therapeutic horseback riding on school-age children and adolescents with Autism Spectrum Disorders. Research in Autism Spectrum Disorders, 6, 578-588.

Horses and Humans Research Foundation (2014a). Article in prestigious journal says EAT should be avoided. Newsletter, 2, 2014. Retrieved from: http://horsesandhumans.org/Newsletters/hhrfnlissue2 2014.pdf

Horses and Humans Research Foundation (2014b). HHRF Webinar Companion, August 2014: Lay language summary. Retrieved from: www.horsesandhumans.org

Kern-Godal, A. (2014). Something good in everything I see. Newsletter - Horses in Education and Therapy International, December 2014, 1 & 3. Retrieved from: http://www.hetifederation.org/pdfs/Newsletters/1214.pdf

McVeigh, T. (2012, February 25). Not just horsing around … psychologists put their faith in equine therapies. The Observer. Retrieved from: http://www.theguardian.com/society/2012/feb/26/horses-therapists-stress-autism-addiction

Nurenberg, J. R., Schleifer, S. J., Shaffer, T.M., Yellin, M., Desai, P.J., Amin, R., Bouchard, A., Montalvo, C. (2015). Animal-assisted therapy with chronic psychiatric inpatients: equine-assisted psychotherapy and aggressive behavior. Psychiatric Services 66(1), 80-86.

Wilkins, C. (2013, September). Healing with the herd. Horse and Rider, pp. 60-64.

Article One

The Effect of Therapeutic Riding on Children with Autism: Two Case Studies

Jessica Johnson, Nicole Nicholson, and Kathryn Potoczak, Shippensburg University of Pennsylvania

Correspondence to:
Dr. Kathryn Potoczak, BCBA-D, Department of Psychology, Shippensburg University of Pennsylvania, 1871 Old Main Drive, Shippensburg, PA 17257
Email: KMPoto@ship.edu

Acknowledgements:
This study was approved by the Shippensburg University Committee on Research on Human Subjects on April 20, 2014 (R# 03-10-14-ASB) and was supported in part by a grant from the Summer Undergraduate Research Experience (SURE). This study would not have been possible without the staff and volunteers at Walnut Grove Therapeutic Center, who contributed countless hours to the project

Abstract

Two case studies examined the impact of therapeutic riding on behavior change in two children with autism. Two children participated in six weekly sessions of therapeutic riding lessons, which were tailored to the specific needs of each child. Communicative and motor behaviors were assessed before and after each session using relevant portions of the Childhood Autism Rating Scale (CARS) and the Vineland Adaptive Behavior Scale (VABS). In addition, direct observations of communicative behaviors were undertaken during each lesson. After six weeks of therapeutic riding, no significant changes were found for the CARS. Both children showed improvement on the VABS with regard to communication and motor skill development. Frequency of clear communications increased in both children across lessons.

Keywords

Autism, Therapeutic Riding, Horse Riding, Communication, Motor Skills.

Introduction

Horses can serve as a source of support for children, especially those with disabilities (Gabriels et al., 2012). Children with autism may often feel socially excluded, but animal-assisted interventions, in particular with equines, may offer an approach to treatment where children can benefit from this particular type of interaction (Gabriels et al. 2012). For a diagnosis of autism to be established, children must have deficits in the areas of communication as well as in social interaction which are observed in a variety of situations. Children with autism also display restrictive and repetitive patterns in behavior, interests, or activities, and the disorder must cause clinically significant distress to the child in social, occupational, or other areas of functioning (American Psychiatric Association, 2013). Although various techniques have been used to achieve behavior change in children with autism, equine-assisted interventions are one new alternative. Animal-assisted interventions offer an approach to treatment that may improve physical ability, motor skills, communication, control, attention, and sensory management (Ward, Whalon, Rusnak, Wendell, & Paschall, 2013). Working with equines may stimulate the development of physiological, psychological, and social skills in children with autism, while providing a unique environment in which the child is able to relax in response to the feeling of the horse's gentle movement (Ward et al., 2013).

The use of the horse as a therapeutic strategy can be especially effective for children with autism as multiple domains (cognitive, social, and physical) may be targeted all at once (Bass, Duchowny, & Llabre, 2009). In their study, Bass et al. (2009) found that in comparison to a wait list group, children with autism who participated in therapeutic riding showed an improvement in sensory integration, directed attention, social motivation, and sensory sensitivity, as well as a decrease in inattention and distractibility. Gabriels et al. (2012) found significant improvements between baseline and post-intervention scores on motor skills for children with autism after engaging in riding lessons. In addition, scores on lethargy, stereotyped behaviors, irritability, and hypersensitivity significantly decreased after equine assisted activities (Gabriels et al., 2012). In her systematic literature review, O'Haire (2012) found that children with autism experienced significant improvement in mood following interactions with animals, as evidenced by increase in smiling, laughing, and positive expressions of emotion.

Lanning, Baier, Ivey-Hatz, Krenek, and Tubbs (2014) used quality of life assessments to determine global changes in functioning in ten children diagnosed with autism spectrum disorder (ASD) who participated in therapeutic riding as compared to eight children with ASD who participated in a non-equine intervention. While all of the children benefitted from intervention, children who participated in therapeutic riding demonstrated greater improvement. Most recently,

Gabriels et al. (2015), in the first large-scale, randomized, controlled trial study on therapeutic riding in children and adolescents with ASD, observed significant improvement in the therapeutic riding group as compared to the control in social cognition and communication, as well as in the total number of words and new words produced.

Therapeutic riding activities involve not only the participant but can also involve the parents of the child. Together, the parent and instructor set goals and determine target behaviors to improve. The instructor can then create physically and mentally-stimulating lesson plans for the child. Holm et al. (2014) tracked the progress of children with autism in which the instructor identified areas for improvement with the parents and child, and the researchers found that 70% of the selected behaviors were improved during four weeks of weekly lessons. Notably, the researchers discovered that improvement in 63% of these behaviors was maintained during reversal phases when treatment was withdrawn. Additionally, some of the effects were generalized to the home environment. Parents in the study noted an increase in their child's vocalizations and ability to follow directions, as well as better physical strength and muscle coordination.

Therapeutic riding is a unique type of intervention because it can provide a peaceful yet still educational alternative to other approaches. Sams, Fortney, and Willenbring (2006) found that children with autism engaged in occupational therapy involving animals showed a significantly greater increase in language production and social interactions than children with autism who were in standard occupational therapy without animals. After three and six months of riding, Kern et al. (2011) found a significant decrease in Childhood Autism Rating Scale (CARS) scores, showing a significant improvement in core autism symptoms from therapeutic riding. The researchers also noted an improvement in mood, sensory processing, and improvements on a quality of life measure. Sams et al., as cited in Koegel & Brookman (2005), explained that when children with autism were allowed to play and learn in a natural environment, researchers could use play and the fun atmosphere to motivate the children to engage in learning the desired target behaviors. These researchers reported that in such an environment, children learned new behaviors faster, showed a decrease in disruptive behavior, and were able to generalize the learned skills into other situations. Therefore, equine assisted activities that take place in a natural environment with various opportunities for stimulation may be motivating for children with autism.

Although much of the previous research has suggested that therapeutic riding can serve as an agent for behavior change in children with autism, there has been a lack of research focused on specific domains of improvement. While some researchers have offered that children with autism have shown an improvement in the CARS after therapeutic riding (Kern et al., 2011), there is a lack of data on the individualized changes and improvements that occur. While there is some evidence that therapeutic riding can be beneficial for children with autism, a study with a focus on specific areas of deficits related to the disorder such as communication and motor skills could illuminate the areas in which children are able to improve after horseback riding lessons. To track individualized progress more thoroughly, a case study approach is required to achieve an in-depth analysis of specific improvements. Research is needed to explore children's progress in motor skill development and communication by incorporating detailed recording of specific behaviors, rather than measuring a broad spectrum of behaviors.

In our case studies, specific behavioral changes that occurred in children with autism as they underwent therapeutic riding were assessed. The goals of the studies were to measure behavior change via motor movement and vocalizations over a period of six weeks. The two children in the study worked with a certified therapeutic riding instructor at an established, insured facility. Lesson plans were created to be cognitively and physically stimulating, while remaining enjoyable for the rider. Before, during, and after six weeks of intervention, motor skills and communication in children with autism were evaluated, as well as overall well-being.

METHOD

Two children diagnosed with autism, one boy (age seven) and one girl (age five) from south central Pennsylvania, were volunteered for the study by their parents in response to advertisements for therapeutic riding for children diagnosed with autism posted at area schools for children with special needs. The diagnoses of autism were assigned by medical professionals independent of and prior to the study, as this served as criteria for inclusion; however, there was no additional information garnered regarding severity. The horses used were a twenty-two-year old Quarter horse mare who had been involved in the lesson program for five years and a twelve-year-old Pony of America breed that had been in the lesson program for seven years. Both horses had a very quiet and willing demeanor.

INSTRUMENTS

CARS AND VABS. The questionnaires used for the study included the portions of the Childhood Autism Rating Scale (CARS) and the Vineland Adaptive Behavior Scale (VABS) used to assess communication and motor skills. The CARS is used to identify children ages two years or older with autism, as well as to determine their level of autism. This ranges on a scale of mild, moderate and severe, and is based in large part on motor behavior and communication. It is for this second area of communication that the CARS was utilized in the present study, and the parents and the riding instructor were asked to fill out this scale. The CARS has been shown to be a reliable and valid measure, and has shown agreement with the DSM-IV (Rellini, Tortolani, Trillo, Carbone & Montecchi, 2004). In addition, due to its good internal consistency (89.6%) and interrater reliability (79.6%), Breidbord and Croudace (2013) recommend its use in early phases of exploratory research.

The Vineland Adaptive Behavior Scale (VABS), the most widely-used adaptive behavior test in the world (Sparrow, 2011), has been demonstrated to be both reliable and valid, and can be used with children at various cognitive levels (Perry, Flanagan, Gier, & Freeman, 2009). Most recently, Yang, Paynter, and Gilmore (in press) used the VABS to develop an adaptive behavior profile for 77 children diagnosed with ASD. The VABS is designed to measure adaptive behavior using four domains: Communication, daily living skills, socialization, and motor skills (Perry, Flanagan, Gier, & Freeman, 2009). The scores provide an adaptive behavior composite that can be used to assess changes in behavior over time. While there are versions of the scale that are to be completed by professionals who interact with the child, such as teachers, instructors, and support staff, the scale used in this study was the version designed for parents to rate their children. As this study focused on only two of the four domains assessed by the VABS, communication and motor skills, parents were asked to complete only those portions of the scale.

Direct observation of communicative behavior. During the lessons, the researchers were located in the center of the arena, in order to observe all interactions between the children, instructor, and horse. After obtaining parental consent to do so, lessons were digitally recorded, in order to more accurately capture communicative behaviors. Upon subsequent viewing of the recordings, researchers assessed the frequency of two types of vocalizations for each child during each lesson: 1) jumbled words, defined as vocalizations lacking identifiable words, and 2) clear communications, which were defined as vocalizations with clear and identifiable words.

Procedure

Before the lessons began, the children were asked to come to the riding facility for an evaluation and for parents to fill out the required paperwork. All lessons were conducted by the same instructor each week, who holds an undergraduate degree in Equine Facilitated Therapeutics and phase two certification through the Pennsylvania Council of Therapeutic Horsemanship. Prior to the first lesson, information regarding height, weight, and personality of the child was taken into consideration in order for the instructor and researchers to pick the most appropriate horse for the child to interact with at his or her evaluation. At the first lesson, the children were asked to give verbal consent as to whether or not they wished to participate in the riding activities. Parents were asked to sign a riding liability form and an informed consent form; they also completed the relevant portions of the CARS and VABS based on their child's current behavior at home. Parents completed the scales during the first evaluation lesson and again after the last lesson. The CARS scale was also filled out by the instructor for each child during the first evaluation lesson and after the last lesson.

Once the paperwork was completed, the children were taken to various locations around the facility so the researchers and instructor could observe their reactions to various animals, textures, and noises. Evaluations lasted approximately one hour for each child. The children were asked during the evaluation to brush the horse, identify various body parts on the animal, and try on helmets. After the evaluation lesson, the researchers and instructor confirmed the horse choice and the children remained on the same horse during weekly lessons for the six weeks. The equines were carefully chosen; the girl with a laid-back, quiet personality who was of small body type was given a pony with gentle, soft movement and a calm disposition, while the boy who needed more sensory stimulation was given a horse who had a larger gait, yet also had a calm disposition so that the animal would not be disturbed by the child's loud noises or frequent movement.

Lesson plans. Individualized lesson plans with specific verbal and motor tasks to be performed while on horseback were created based on the observed level of functioning for each child during the initial evaluation, and new plans were created for each subsequent lesson after the instructor and researchers reviewed notes and observations from the previous lesson. Within the riding arena, stations with stimulus materials such as different colored balls were set up before the lesson. Upon arrival at the station, the instructor held up a specific colored ball and asked the child to say what color it was. Correct answers were reinforced with praise; incorrect answers were corrected and the child was asked to try again before moving on to the next station or practicing more riding techniques. All lesson plans were reviewed by the instructor and researchers before lessons began. Two side aides and one horse leader were trained on safety techniques and briefed on the nature of the study before helping with lessons. Before the lessons began, the instructor and researchers informed the side aides and leader on the lesson plan so they were aware of the lesson format and could prepare the arena. All lessons were conducted in the outdoor arena. The side aides were asked to refrain from speaking during sessions, in order to track communication from the children in the clearest fashion possible as well as to increase the safety of the rider. When the children arrived at the facility, they were asked to meet the instructor down at the riding ring where their horse was prepared with the correct saddle and riding equipment, ready for the lesson. Right before getting on the horse, the children were required to place a riding helmet on their head and have it secured as a safety precaution.

Mounting and warm-up. Each horse was lined up next to a large mounting block outside the ring where the child could easily place his or her foot in the stirrup and swing his or her leg over the horse to sit comfortably in the saddle. For the girl, a bareback pad was found to be more effective than a saddle after the first lesson, so the pad replaced the saddle for the remaining lessons. The decision for a bareback pad was made because the instructor believed the child needed more stimulation and closer contact with the horse. After mounting the horse outside the ring and being introduced to the side aides for the day, the children were led into the riding ring by the horse leaders with additional side aides on each side. The researchers observed from the middle of the ring, marking data on communicative behaviors as the lesson progressed as well as videotaping the lesson. The children became oriented with their surroundings during a warm-up session where the leader walked the horse around the ring as the riding instructor asked questions and worked to address the children's postures. This portion of the lesson allowed the horses to warm up and the children to focus on the riding instructor. After this, each lesson followed the individualized plan created by the riding instructor. Lessons lasted no more than twenty-five minutes and ended when the child had successfully completed the lesson goals.

RIDING SKILLS. Individualized lesson plans were created to stimulate physical movement and verbal communication. For both children, lesson plans focused on developing the child's clear verbal communication to the horse as he or she was progressively taught and asked to tell the horse when to walk, trot, and stop. For the girl, the six weeks of lessons focused on communication to the horse and some motor movements such as throwing a ball or picking up a ring. The girl was also asked to name the colors of objects as she picked them up and/or threw them. The games both children played focused on physical movement and verbal communication, but were also created to stimulate the children and to help keep their focus on the lesson. For the boy, who was able to grasp more difficult tasks but had difficulty following multiple directions at once, the instructor encouraged more complicated tasks and worked to teach the child how to hold reins, steer the horse, and pull back on the reins while saying "whoa" to halt the horse. Over time, the boy's lesson plans consisted of games as well as other activities that emphasized following multiple directions. Throughout instruction, as the children completed each task properly, they were positively reinforced with praise and by activities they enjoyed, such as being allowed to trot the horse. This served as a reinforcer for the two children because they liked the faster movement of the horse.

COOL-DOWN. Near the end of each lesson, the children were given various options for a cool-down. Both children were given the opportunity to walk to the creek and ride outside the ring; however, only the boy was interested in riding on the trail. The girl would dismount to indicate she was done with her lesson and she seemed to favor visiting the ducks, one of her favorite animals on the farm, as soon as the lesson was over. The boy was given the option of turning barrels over and collecting the toys inside, which he thoroughly enjoyed. He was also fond of stickers and was reinforced with a sticker right before he left the facility. After walking with the horse back to the barn, children were asked to help brush the horse, and when they felt they were done, they were allowed to leave the facility

Results

For the girl, referred to as Case Study One, no changes were observed between before and after scores from the parent on the relating to people and body use domains of the CARS. Adaptation to change scores decreased from 2.5 (between mildly and moderately abnormal adaptation) to 1.5 (between age appropriate response and mildly abnormal adaptation) and verbal communication scores improved from 2.5 (between mildly abnormal and moderately abnormal communication) to 2 (mildly abnormal verbal communication). Listening response scores regressed from 2 (mildly abnormal listening) to 2.5 (between mildly and moderately abnormal listening). On the instructor's feedback, improvement was shown between the before and after scores on relating to people. The change improved from 3 (moderately abnormal relationships) to 2 (mildly abnormal relationships). However, adaptation to change regressed from 1.5 to 2.5, and the body use, listening response, and verbal communication domains did not change. On the VABS, scores within all domains except talking (regressed by 2 points) and using large muscles (regressed by 2 points) improved or stayed constant upon completion of six weeks of therapeutic riding. Listening and understanding stayed constant in points, and using small muscles increased by 21 points, which is a clinically significant change (Table 1).

For Case Study Two, involving the boy, no change was shown between before and after scores from the parent or instructor within the domains studied through CARS. Scores within all domains of VABS improved upon completion of six weeks of therapeutic riding. Listening and understanding improved by 9 points, talking improved by 9 points, using large muscles improved by 1 point, and using small muscles improved by 11 points (Table 1).

Both children showed an increase in observed vocalizations throughout the course of the weekly lessons (Figure 1). For Case Study One, the frequency of jumbled words decreased throughout the six weeks, while clear communications increased. For Case Study Two, no jumbled words were observed during lessons, and an increase in clear communications was observed.

Discussion

The goal of the two case studies was to see if six weeks of therapeutic riding could generate increases in appropriate communication as well as an improvement in motor skills in children diagnosed with autism. Important to note is that lessons were not run on a half hour or hour basis, but were finished when the child reached a degree of progress that the instructor viewed as improvement. After the lesson, he or she was allowed to participate in a favorite farm activity as a reinforcer for desired behavior.

There were not significant changes observed on the CARS. It may take longer than six weeks to see significant changes regarding the overall severity of the behaviors characteristic of autism. However, both children did show positive changes on the VABS through improvement in the rating of communication and motor skills by both the parent and instructor. This was also reflected in the observation data, as there was an increasing trend in clear communications across lessons.

For Case Study One, the lessons were targeted to improve communication and two-step directions. This was developed in support of research conducted by Ward et al. (2013), in which the researchers note that therapeutic riding can serve as a useful tool for children diagnosed with autism, as the child must focus on control, attention, focus, sensory management, and non-verbal as well as verbal communication in order to engage in riding activities.

The parent reported that her child had improved with regard to following directions at home, so some concepts learned during riding lessons seemed to generalize into the child's home life.

Physical ability improved for the girl. In the beginning of the lessons, the child needed more support from the side aides and instructor to stay balanced; however, during the last lesson it was observed that the child could balance herself and did not need support from the helpers as much as in previous lessons.

For Case Study Two, the child was able to ride without help from the side aides or leaders while steering at the walk. He was able to complete balancing exercises, and showed he was able to ride without holding onto his reins or saddle horn. During the last lesson, he learned how to post during trot. The child had improved in balance, coordination, and muscle strength. His parent informed us that her child was never as calm and relaxed as he was during horseback riding lessons, and that the riding had a positive effect on him. In fact, she stated that she wished he could do everything while on horseback.

Not only was riding an educational opportunity, but it was also a pleasurable experience for the children. Many instances of the children smiling and laughing were observed, especially as they were allowed to trot the horse. The relaxing environment with the opportunity to interact with an animal was a motivation for learning in itself; the horse was used to teach the children to communicate in more effective ways. The parent of the girl reported that her child often mentioned going to see the horses and was very excited each time she was able to return.

As the lessons progressed, the children vocalized more, as they had to speak in order to command the horse to stop, walk, or trot. With each lesson, the riding period became longer, and the children's focus seemed to be sustained longer.

Limitations

Although there is evidence that suggests therapeutic riding had a positive effect on the children in this study, it would have been beneficial to continue the lessons for a longer period of time to see if even greater improvements in the areas of verbal communication and motor skills could be attained. In addition, this may have led to an improvement in ratings of severity of autistic behaviors by parents as assessed by the CARS. Further, it is hard to generalize the results of only two children's experiences to all children with ASD.

Conclusion

Future research should include a study with larger sample size and observations over a longer period of time, as well as additional direct measures of behavior before, during and after therapy. In spite of these limitations, this study further supports previous findings that therapeutic riding does indeed have a beneficial effect on the behavioral deficits commonly observed in children diagnosed with autism. Therapeutic riding should be considered as an alternative or addition to more traditional treatment options as it offers potential for increased enjoyment during the learning experience, as well as the opportunity for extensive parental involvement in both the creation of individual-tailored lesson plans and the experience itself.

APPENDIX A

TABLE 1: SUMMARY OF CARS AND VABS EVALUATIONS

	Case Study One- Initial Evaluation	Case Study One- Final Evaluation	Case Study One- Difference in Evaluations	Case Study Two- Initial Evaluation	Case Study Two- Final Evaluation	Case Study Two- Difference in Evaluations
CARS Rating- Parent						
- Relating to people	2.5	2.5	0	2	2	0
- Body use	2	2	0	2	2	0
- Adaptation to Change	1.5	2.5	1	2	2	0
- Verbal Communication	2.5	2	-0.5	2	2	0
- Listen Response	2	2.5	0.5	3	3	0
CARS Rating- Instructor						
- Relating to people	3	2	-1	1	1	0
- Body use	1	1	0	1	1	0
- Adaptation to Change	1.5	2.5	1	1	1	0
- Verbal Communication	2	2	0	1	1	0
- Listen Response	2	2	0	1	1	0

TABLE 1: SUMMARY OF CARS AND VABS EVALUATIONS – CONT'D.

	Case Study One- Initial Evaluation	Case Study One- Final Evaluation	Case Study One- Difference in Evaluations	Case Study Two- Initial Evaluation	Case Study Two- Final Evaluation	Case Study Two- Difference in Evaluations
VABS Rating- Parent						
- Listening and Understanding	**22**	**22**	**0**	**19**	**28**	**9**
- Using Large Muscles	**35**	**33**	**-2**	**17**	**18**	**1**
- Using Small Muscles	**9**	**30**	**21**	**22**	**33**	**10**
- Talking	**40**	**38**	**-2**	**34**	**43**	**9**

Table 1: Summary of CARS and VABS evaluations. This table shows a summary of the CARS data collected from both the parent and the instructor during the initial evaluation and the final evaluation. It also summarizes the VABS data collected from the parents during the initial evaluation and the final evaluation.

APPENDIX B

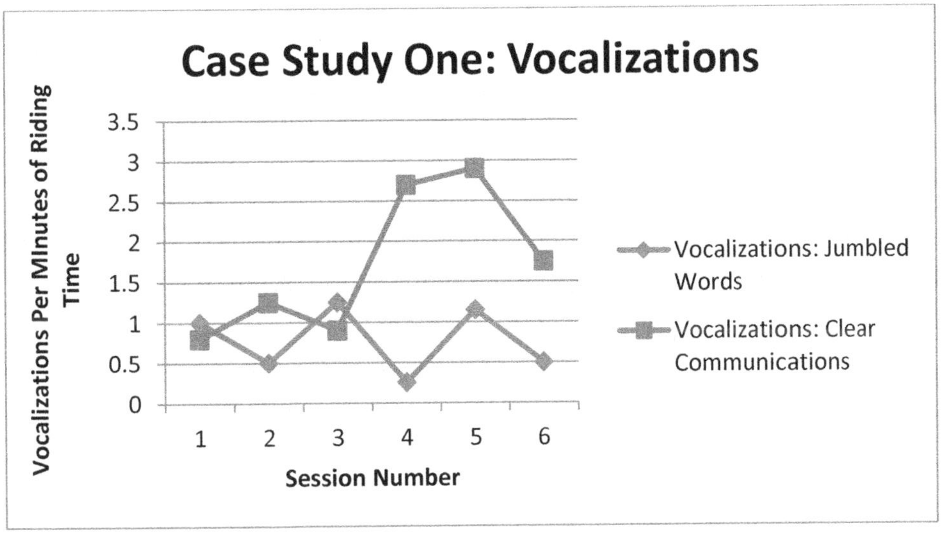

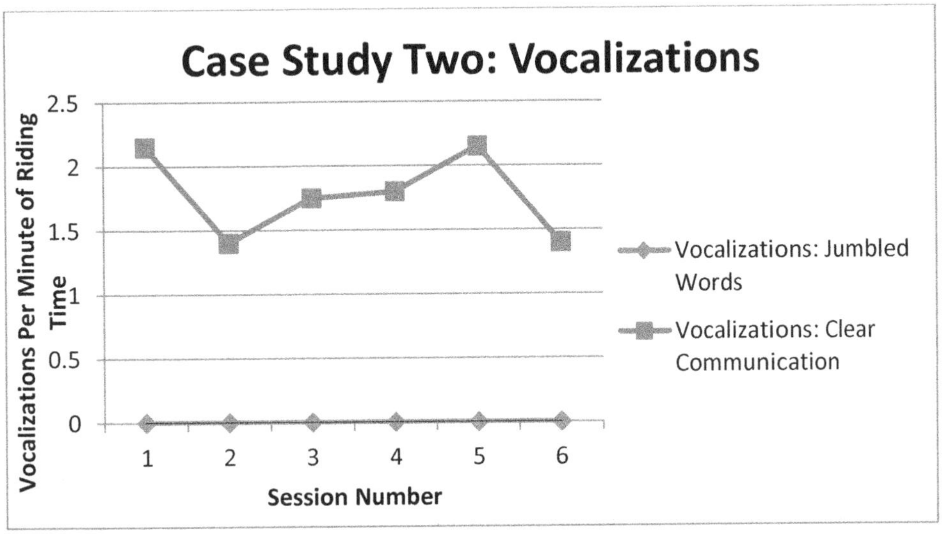

Figure 1: Vocalizations from Case Study One and Case Study Two. These graphs visually represent the change in frequency and type of vocalizations over the course of the six-week intervention.

REFERENCES

American Psychiatric Association. (2013). Diagnostic and statistical manual of mental disorders (5th ed.). Washington, DC: Author.

Bass, M. M., Duchowny, C. A., & Llabre, M. M. (2009). The effect of therapeutic riding on social functioning in children with autism. Journal of Autism and Developmental Disorders, 39, 1261-1267.

Breidbord, J. & Croudace, T. J. (2013). Reliability generalization for Childhood Autism Rating Scale. Journal of Autism and Developmental Disorders, 43(12), 2855-2865.

Gabriels, R. L., Agnew, J. A., Holt, K. D., Shoffner, A., Zhaoxing, P., Ruzzano, S., Clayton, G. H., & Mesibov, G. (2012). Pilot study measuring the effects of therapeutic horseback riding on school-age children and adolescents with autism spectrum disorders. Research in Autism Spectrum Disorders, 6(2), 578-588.

Gabriels, R. L., Pan, Z., Dechant, B., Agnew, J. A., Brim, N., & Mesibov, G. (2015). Randomized controlled trial of therapeutic horseback riding in children and adolescents with autism spectrum disorder. Journal of the American Academy of Child and Adolescent Psychiatry, 54(7), 541-549.

Holm, M. B., Baird, J. M., Kim, Y. J., Rajora, K. B., D'Silva, D. D., Podolinsky, L., Mazefsky, C., Minshew, N. (2014). Therapeutic horseback riding outcomes of parent-identified goals for children with autism spectrum disorder: An ABA multiple case design examining dosing and generalization to the home and community. Journal of Autism & Developmental Disorders, 44, 937-947.

Kern, J. K., Fletcher, C. L., Garver, C. R., Mehta, J. A., Grannermann, B. D., Knox, K. R., Richardson, T. A., & Trivedi. (2011). Prospective trial of equine-assisted activities in autism spectrum disorder. Alternative Therapies, 17(3), 14-20.

Lanning, B.A., Baier, M.E., Ivey-Hatz, J., Krenek, N., & Tubbs, J.D. (2014). Effects of equine assisted activities on autism spectrum disorder. Journal of Autism & Developmental Disorders, 44(8), 1897-1907.

O'Haire, M. (2013). Animal-assisted intervention for autism spectrum disorder: A systematic literature review. Journal of Autism & Developmental Disorders, 43(7), 1606-1622.

Perry, P., Flanagan, E. F., Gier, J, D., & Freeman, N, L. (2009). Brief report: The Vineland Adaptive Behavior Scales in young children with autism spectrum disorders at different cognitive levels. Journal of Autism & Developmental Disorders, 39(7), 1066-1078.

Rellini, E., Tortolani, D., Trillo, S., Carbone, S., & Montecchi, F. (2004). Childhood Autism Rating Scale (CARS) and Autism Behavior Checklist (ABC) correspondence and conflicts with DSM-IV criteria in diagnosis of autism. Journal of Autism and Developmental Disorders, 34(6), 703-708.

Sams, M. J., Fortney, E. V., & Willenbring, S. (2006). Occupational therapy incorporating animals for children with autism: A pilot investigation. American Occupational Therapy Association, 60, 268–274.

Sparrow, S. S. (2011). Vineland Adaptive Behavior Scales. In Kreutzer, J. S., DeLuca, J., & Caplan, B. (Eds.), Encyclopedia of clinical neuropsychology (pp. 2681-2621). New York: Springer.

Ward, S. C., Whalon, K., Rusnak, K., Wendell, K., & Paschall, N. (2013). The association between therapeutic horseback riding and the social communication and sensory reactions of children with autism. Journal of Autism and Developmental Disorders, 43, 2190-2198.

Yang, S., Paynter, J. M., & Gilmore, L. (in press). Vineland Adaptive Behavior Scales: II profile of young children with Autism Spectrum Disorder. Journal of Autism and Developmental Disorders.

ARTICLE TWO

Impact of Therapeutic Riding for Families with ASD children: A Case Study

Veronica Lac, MA, A True Image Consulting Inc., Saybrook University PhD (Psychology) Candidate June 2015

Correspondence to:
Veronica Lac,
5195 Hampsted Village Center Way, Unit 59
New Albany, Ohio, 43054
Email: veronicalac@hotmail.com

ABSTRACT

Autism spectrum disorders (ASD) are associated with a wide range of functional impairments that can cause multiple challenges to the daily lives of affected families. The impact of therapeutic riding lessons on families with children diagnosed with ASD is discussed in this paper. Through the exploration of the key stressors on these families, a case study is used to illustrate the benefits of therapeutic riding for the whole family. Therapeutic riding engages these children cognitively, emotionally, and physically and has the potential to improve their social and verbal skills as well as increase their sensory processing tolerance. This, in turn, improves their ability to tolerate changes in their routine, communicate within their family, and improve parent-child and sibling relationships whilst increasing the family's sense of freedom.

KEYWORDS

Therapeutic Riding; Family Impact; Autism; Sibling Relationships.

INTRODUCTION

Autism spectrum disorders (ASD) represent a range of neurological developmental disorders that affect fundamental aspects of personhood very early on in the developmental process (McGrath & Peterson, 2009). ASD is categorized under the umbrella of pervasive developmental disorders (PDD), and is associated with impaired social communication, sensory integration, motor skill functioning, and the presence of restricted and/or repetitive patterns of behavior (DSM-5, 2013). There are three levels of severity based on the amount of support required, and the symptoms may or may not be accompanied by intellectual or language impairment, or association with any medical or genetic conditions or environmental factors.

This multifaceted and pervasive nature is reflected in the effects on the parents and families of a child with ASD (Karst & Van Hecke, 2012). Receiving an ASD diagnosis for their child often result in lost hopes and dreams for parents that lead to personal and professional sacrifices, increased stress and conflicts within marital relationships that impact the parent-child relationship, and changes in family dynamics (Nealy, O'Hare, Powers & Swick, 2012).

In this paper, these main stressors that families of children with ASD face are demonstrated through the use of a case study. Also discussed is how therapeutic riding lessons can help to alleviate some of these challenges.

KEY CHALLENGES FOR FAMILIES OF CHILDREN WITH ASD

The process from initial diagnosis and intervention planning through to the acceptance of the realities of raising a child with ASD represent an incredibly stressful, emotional, and logistically challenging journey for many families (Bloch & Weinstein, 2010). Many have to adjust their employment status and thereby increasing financial pressures. Raising a child with ASD also impacts the parent-child relationship, as well as increasing marital and sibling relationship problems (Karst & Van Hecke, 2012).

LOST DREAMS AND RELATIONSHIPS

With an ASD diagnosis, parents often experience a deep sense of loss and grief for the hopes and dreams that they had for their child (Nealy et al., 2012). This may be in the form of expectations of their child to follow a certain career path, or hopes for marriage and grandchildren, as well as hopes of having a happy, healthy, "normal" child. Particularly for parents of children who are non-verbal and requiring more assistance, the realities and worries of how they might be supported when the child is older can increase propensity of parental depression

(Karst & Van Hecke, 2012). Additionally, parents are often met with a lack of support from their wider families and social circles that may lead to loss in friendships (Nealy et al., 2012), which may further exacerbate feelings of isolation and loss. Furthermore, professional opportunities may be lost through the need to maintain consistency for their child, and career goals may be sacrificed (Bloch & Weinstein, 2010).

Rigid Routines & Meltdowns

Bloch and Weinstein (2010) asserted that one of the most challenging aspects of raising a child with an ASD diagnosis is the invisibility of the child's disability. The neurological impairments that underlie the sensory and self-regulatory issues present families with a range of difficulties in daily life. From public meltdowns due to sensory overload, to unusual eating, sleeping or dressing patterns, and the child's excessive need for routine, these challenges affect the whole family. "A simple drive to a store might trigger severe tantrums because an adult needed to change their route…a sibling's wish to change a channel on a television can be enough to precipitate a major reaction (Bloch & Weinstein, 2010, p. 33).

Many parents struggle in public with judgmental comments from passers by, who assume that these meltdowns are due to these children being "spoilt", triggering feelings of embarrassment, shame, and hurt (Cassidy et al., 2008) which may further increase their sense of isolation.

Affective Reciprocity & Family Dynamics

Children with ASD are often unable to demonstrate affective reciprocity (Karst & Van Hecke, 2012). Not only are smiles, eye contact, and affectionate touching and hugging that are initiated by family members often not reciprocated, but may become triggers for sensory overload and meltdowns. Karst and Van Hecke (2012) suggested that this lack of affective reciprocity is more damaging to family dynamics than the cognitive or behavioral challenges associated in caring for a child with ASD.

Family dynamics are challenged when a child is diagnosed with ASD due to increased stress related to financial and time pressures on the parents, conflicts between spouses on behavioral issues and treatment options, as well as decreased parental attention for siblings (Karst & Van Hecke, 2012). Furthermore, sibling relationships may suffer due to behavioral issues displayed in public and/or parental expectations for siblings to be responsible for the child with ASD. Siblings may express their embarrassment at inappropriate behavior, and feelings of being burdened by the responsibility of their ASD sibling. They may also feel the need to overachieve in their own life to compensate for their sibling's disabilities (Bloch & Weinstein, 2010).

For children who are non-verbal, the lack of affective reciprocity combined with the lack of verbal communication often leaves family members at a loss of how to engage with the child that results in increased frustration for all involved and impacts the overall family dynamics as well as parent-child and sibling relationships (Karst & Van Hecke, 2012).

Benefits of Therapeutic Riding

Therapeutic riding, which incorporates recreational activities involving riding and horsemanship for individuals of all abilities, offers an embodied multisensory experience that affects the client's sensory, neurological, psychological, and social processing patterns (Cuypers et al., 2011). Riders are encouraged to use verbal commands as well as body cues that incorporate the movement of seat, leg, and hands (PATH International, 2015). Horseback riding improves balance and core strength, as well as fitness and endurance, as sitting on the horse and staying balanced requires the use of core muscles (King, 2007). At different gaits, the horse provides different movement patterns and sensory input to the rider (Trotter, 2012).

There is support that therapeutic riding is beneficial for individuals diagnosed with ASD (Ward et al., 2013; Gabriels et al., 2012; Bass et al., 2009). In particular, Gabriels et al. (2012) found in their pilot study that children attending therapeutic riding showed significant improvements on measures of self-regulation, adaptive expressive language skills, motor skills, and verbal praxis/motor planning skills after as few as 10 weekly lessons.

Bass et al. (2009) focused on the impact of therapeutic riding on sensory integration, which is understood as the processing of external stimulus into corresponding motor and behavioral responses (SPDF, 2014). The results showed a positive improvement in sensory integration, directed attention, social motivation, and sensory sensitivity. Holm et al. (2014) indicated that the most marked change in behaviors following participation in therapeutic riding was that of spontaneous verbalization, which occurred both during the lessons as well as at home and at school. Ward et al. (2013) researched the association between therapeutic riding and the social communication and sensory processing skills of elementary school children with ASD, and concluded that the positive effects were long lasting enough to be carried through to classroom environments, with improved social communication, attention and sensory reactions.

Case study

For the purposes of this discussion, the personal, identifying details of my client have been changed in order to maintain confidentiality. Informed consent was obtained from the client's family for inclusion in this case study.

Context of Practice

I am a Registered Therapeutic Riding Instructor, certified through PATH International in the USA. Additionally, I am also a Doctoral Candidate in a PhD Psychology program, and a Licensed Professional Counselor.

As a PATH International Therapeutic Riding Instructor, I conduct lessons in a small group format of no more than three participants per lesson, with children ranging in age from 4 and 14 years. The majority of these participants have been diagnosed with ASD and many of them have dual diagnoses that include anxiety disorders, obsessive-compulsive disorder, and Attention Deficit Hyperactivity Disorder. The therapeutic riding program is based in Ohio, USA, and has been in operation for 4 years. Within the program, I am also responsible for the lesson planning and outcome evaluations, volunteer training and development, and intake assessments for special needs clients. Our therapeutic riding lessons rely on the support of volunteers who provide the critical roles of horse leaders and side-walkers.

Assessment Session

Before introducing a child into the group lesson format, an initial assessment is conducted to ascertain the levels of support they might need during the lessons. The assessment consists of an interview with the parent(s) or legal guardian and an evaluation of the child's behaviors and physical ability. The evaluation, based on the guidelines provided as part of the certification process for PATH registration (PATH 2015), provides a superficial glance at the child's gross motor co-ordination and balance, as well as their ability to follow instructions. For example, their ability to hold a ball in each hand for any length of time is a useful gauge to know whether they might be able to hold their reins whilst riding; and their ability to stand on one leg can indicate whether they need additional support to mount the horse. Additionally, what type of ball they choose might indicate a preference for a particular texture, firmness or color, and from a psychological perspective, how long they hold onto the item might also indicate the need for a transitional object during lessons (Rector, 2005). These are small but significant choices that the rider can make without guidance and they are particularly important when dealing with children who are non-verbal and/or have sensory processing challenges. The assessment also includes a brief interaction with one of the horses and a trial helmet fitting process. This provides a glimpse into the tolerance level for the different sensory inputs involved in wearing a helmet and in being with horses.

Amy

This case study is based on the observations from 2 years of working with the client, Amy. During the time span of the case study, Amy was also engaged in Applied Behavior Analysis (ABA) therapy and speech therapy. Her mother had been referred to our therapeutic riding program by their pediatrician.

Amy arrived at her assessment session in her mother's arms, hidden under a blanket. She was 5 years old and had been diagnosed with ASD by her pediatrician six months prior to our meeting. Amy displayed many of the characteristic impairments of ASD, including repetitive and obsessive-compulsive behaviors and the need for precision routines. She was non-verbal, had sensory processing difficulties, and had trouble following instructions. Her gross and fine motor skills were typical for her age. Amy had two siblings, an older sister and a younger brother. Her mother explained that she felt that the other two children did not receive the attention they deserved, and that this was a constant worry for her. Her mother also indicated that due to Amy's obsessive-compulsive behaviors, the whole family became subjected to a military precision routine that centered on her preferences. The introduction of anything new in her routine would result in "screaming meltdowns" that could last for hours, so they spent their lives trying to avoid triggering them. Her mother had recently resigned from her full-time position in order to manage the daily schedule of therapy visits that Amy required with the responsibilities of raising two other children. Amy's father had taken an extra part-time job in addition to his full-time position to compensate for some of the lost income.

During the assessment session, Amy was invited to choose between toys of various shapes, sizes, textures, and colors. Her consistent choices included a small, soft teddy bear and a medium sized, smooth red rubber ball. Helmet fitting was challenging as she did not like to be touched under her chin for the strap to be fastened, but with encouragement, became calm enough for the helmet to be secured. When she was introduced to the horses, Amy began to smile, laugh and stretched out her hand to pet them. For the first time during the assessment, she voluntarily left her mother's side and positioned herself next to Duke, one of the therapy horses, hugged his leg and buried her face into his shoulder. Duke, meanwhile, turned towards her and wrapped his neck around her shoulders. This sent Amy into squeals of delight that brought her mother to tears as she explained that this was the first time that she had witnessed her daughter's ability to experience joy.

Establishing routines whilst introducing change

Because of her good balancing and gross motor co-ordination skills, I had assigned Amy a team of two volunteers consisting of a horse leader and one side-walker. The side-walker was responsible for Amy's safety both on the ground and whilst mounted, and was briefed on Amy's obsessive-compulsive behaviors. I placed Amy in a group with two other non-verbal children of approximately the same age who had also been diagnosed with ASD. Additionally, I attempted, whenever possible, to maintain the same horse leader and side-walker each week, as well as reserving the horse, Duke, who she had connected with during her assessment as her lesson horse.

Since all three participants in this group were non-verbal and also struggled with sensory processing, I began each lesson by asking them to groom their horses. This provided them some adjustment time to transition between their previous surroundings (e.g., school, home, or therapist office), to being at the barn with the horses. It also provided them with the opportunity to focus their sensory awareness at a more grounded pace compared to the sensory arousal of being mounted on a horse. Over the course of the first few lessons, I noticed that Amy would remember everything about the grooming process; from the particular brush that was used to brush Duke, to the color of the lead-rope and halter that he had. If anything changed from one week to the next, she would have a meltdown. Each time this occurred, I would redirect her attention to what was consistent and give her the soft teddy bear or the smooth red rubber ball to hold, which would calm her down enough for us to continue with the lesson.

THE POWER OF SPEECH

Each week, I designed lesson plans and goals based on the participants' physical abilities such as stopping their horses at a stop sign, or standing in their stirrups unaided while the horse walks over ground poles in the arena, or holding onto their reins and learning the beginning elements of how to steer their horses. These skills were broken down into sequential instructions that were then taught one at a time. Activities were also incorporated into the lessons to make it more interesting and engaging for the participants. The lessons were designed to be inclusive for both verbal and non-verbal presentations, in the hope that during the process of learning, the children would develop verbal skills. Side-walkers were trained to encourage participants to use their verbal skills as well as incorporating non-verbal cues into the process. For example, in learning to stop, participants were asked to cue their horses by pulling back gently on both reins, leaning back slightly in their saddle, and saying the word 'whoa'. Amy was able to use her reins and lean back, but instead of saying 'whoa', she would let go of her reins and use sign language to say 'stop'. Similarly, when asking the horse to 'walk on', instead of verbalizing this, Amy would tap her hand on the front of her saddle to indicate her intention for the horse to walk.

After two months of weekly, 45-minute group lessons, Amy stood at the mounting block patiently waiting to mount her horse. Once mounted, her side-walker turned to her and asked her to signal the horse to walk on with a tap on the saddle. Instead, Amy looked up and said, "Walk". This was her first spoken word. Two weeks later, her mother reported that Amy walked into her bedroom that morning and said, "Hi, Mommy".

Progress

Amy has now been part of the therapeutic program for two years. Her riding skills have progressed significantly and she is now able to walk and trot without being held by her side-walker, and is learning the basics of steering. Amy is currently in the process of preparing for her first Special Olympics competition and is determined to win a trophy. During lessons, her spontaneous verbalization has increased and she is now able to say short sentences. One of the lesson games involves steering the horses to kick a giant soccer ball across the arena. This is particularly useful in supporting Amy to verbalize and build on her sentence structure from simply asking Duke to "kick", or saying "ball", to "Duke, kick ball", and progressing to "Duke is kicking the ball."

Amy is now also able to tolerate changes in routines more, although she still requires her volunteer team to be as consistent as possible as her ability to focus on tasks during lessons is noticeably decreased when her regular team is not available. Her attention span has also increased as well as her ability to follow sequential instructions. Additionally, Amy has made improvements in her ability to maintain eye contact and in her social communication skills, spontaneously hugging her horse and her volunteer team, as well as saying 'thank you' to each of them at the end of every lesson. The affections shown during the therapeutic riding lessons by her horse and volunteer team are now occasionally reciprocated.

These improvements are also seen in the family environment. Amy is now more able to communicate with her family through her power of speech. Although family life is still restricted somewhat to Amy's preferences, it is now possible to negotiate options that provide the family a little freedom from the previously rigid routines. They are able to introduce some variety into their lives in terms of places frequented and people they socialize with. Amy's relationships with her siblings have also improved through this relatively increased flexibility, as well as in her willingness and ability to reciprocate affection. Her increased attention span and ability to follow sequential instructions also enables family life to function more smoothly.

Discussion

Improvements in all the identified key stressors on the family, lost dreams and relationships, rigid routines and meltdowns, and challenges with family dynamics and affective reciprocity, can be seen in this case study. Whilst these changes cannot be wholly attributed to Amy's involvement in the therapeutic riding program, as she has continued with her ABA and speech therapy, they represent significant steps for Amy. Through her contact with horses, volunteers and other participants, Amy's increased tolerance for sensory input and changes in routine allow her and her family more choices in their daily lives. With fewer meltdowns, they are able to take more risks to venture out as a family into the community; her parents and siblings have been able to regain lost friendships, increase social connections, and her mother has been able to start working again part-time. As Amy begins to articulate more clearly how she feels and what she needs, her relationships with both her parents and siblings are less fraught with tension. Amy's affective reciprocity with the horses has transferred to her family environment and she is more willing and able to initiate affectionate touch and hugs, as well as include social pleasantries into her daily life. Amy's participation in the Special Olympics this year provides an opportunity for her parents to discover new dreams of what their child can achieve, and regain a sense of hope for her future.

REFERENCES

Bass, M.M., Duchowny, C.A., & Llabre, M.M. (2009). The effect of therapeutic horseback riding on social functioning in children with autism. Journal of Autism Developmental Disorders, 39, 1261-1267. DOI: 10.1007/s10803-009-0734-3.

Bloch, J.S. & Weinstein, J.D. (2009). Families of young children with autism. Social Work in Mental Health, 8(1), 23-40, DOI: 10.1080/15332980902932342.

Cassidy. A., McConkey, R., Truesdale-Kennedy, M., & Slevin, E. (2008). Preschoolers with autism spectrum disorders: The impact on families and the supports available to them. Early Child Development and Care, 178(2), 115-128, DOI: 10.1080/03004430701491721.

Cuypers, K., De Ridder, K., & Strandheim, A. (2011). The effect of therapeutic horseback riding on 5 children with attention deficit hyperactivity disorder: A pilot study. The Journal of Alternative and Complimentary Medicine, 17(10), 901-908. DOI: 10.1089/acm.2010.0547

DSM-V (2013). Diagnostic and statistical manual of mental disorders, Washington, DC: American Psychiatric Association.

Gabriels, R.L., Agnew, J.A., Holt, K.D., Shoffner, A., Pan, Z., Ruzzano, S., Clayton, G.H., & Mesibov, G. (2012). Pilot study measuring the effects of therapeutic horseback riding on school-age children and adolescents with autism spectrum disorders. Research in Autism Spectrum Disorders, 6, 578-588.

Holm, M. B., Baird, J. M., Young, J. K., Rajora, K. B., D'Silva, D., Podolinsky, L., Mazefsky, C., & Minshew, N. (2014). Therapeutic horseback riding outcomes of parent-identified goals for children with autism spectrum disorder: An ABA multiple case design examining dosing and generalization to the home and community. Journal of Autism and Developmental Disorders, 44(4), 937-947, DOI: 10.1007/s10803-013-1949-x.

Karst, J.S. & Van Hecke, A.M. (2012). Parent and family impact of autism spectrum disorders: A review and proposed model for intervention evaluation. Clinical Child & Family Psychology Review, 15, 247-277, DOI: 10.1007/s10567-0120119-6.

King, N. (2007). Perceived efficacy of therapeutic riding for children with autism. In B. Engel & J. MacKinnon (Eds.), Enhancing human occupation through hippotherapy (pp. 119-126). Bethesda, MD: American Occupational Therapy Association.

McGrath, L.M., & Peterson, R.L. (2009). Autism spectrum disorder, in B.F. Pennington, (Ed.), Diagnosing learning disorders: A neuropsychological framework, (2nd edition), New York, NY: Guildford Press.

Nealy, C.E., O'Hare, L., Powers, J.D., & Swick, D. C. (2012). The impact of autism spectrum disorders on the family: A qualitative study of mother's perspectives, Journal of Family Social Work, 15(3), 187-201, DOI: 10.1080/10522158.2012.675624.

PATH International (2015). www.pathintl.org

Rector, B.K., (2005). Adventures in awareness: Learning with the help of horses. Bloomington, Indiana, USA: AuthorHouse.

SPDF (2015). About Sensory Processing Disorder. Retrieved March 10th 2015 from: http://spdfoundation.net/about-sensory-processing-disorder.html

Trotter, K.S. (2012). Equine assisted interventions in mental health. In K.S. Trotter (Ed.). Harnessing the power of equine assisted counseling: Adding animal assisted therapy to your practice. pp. 1-15, New York, NY: Routledge.

Ward, S.C., Whalon, K., Rusnak, K., Wendell, K., & Paschall, N. (2013). The association between therapeutic horseback riding and the social communication and sensory reactions of children with autism. Journal of Autism and Developmental Disorders, 43, 2190-2198. DOI: 10.1007/s10803-013-1773-3.

Article Three

The Potential of the Horse in Adolescent Aggression Management: A Review of the Literature

Inka Tonteri, M.Soc.Sc., The Finnish Association of Equine Facilitated Therapies, Ypäjä Equine College

Correspondence to:
Inka Tonteri
Email: inka.tonteri@jamsannuorisokoti.fi
Phone: +358503316376

ABSTRACT

Aggression is a natural part of living but aggressive behavior is often a sign of inadequate skills emotion management and insufficient social skills. Aggression education, psychosocial education about aggression, is connected to emotional skills education and psychosocial growth.

The purpose of this study was to explore how equine-assisted activities and therapies (EAAT) can be incorporated into aggression education with an emphasis on children and adolescents. Literature specifically about aggression education and equine assisted activities and therapies is scarce. A secondary purpose was to create a dialog between different scientific theories and equine-assisted work.

Results of this exploration indicate that equine-assisted work as a form of aggression education has many potential elements. These elements are: positive physical sensations and body awareness; synchrony of being and moving; motivation; empathy training; encountering feelings of disappointment and overcoming them; forming a potential space; creating positive interactions and feelings of connectedness. Very few therapeutic or educational interventions can combine these psychosocial and psychomotoric elements.

KEYWORDS

Adolescence, Aggression, Equine-Assisted Activities, Equine Therapy, Aggression Education, Emotional Skills Training.

INTRODUCTION

Aggression is an essential and biological part of being a human and having emotions. Aggression is also reaction, and how it is expressed depends on the individual's childhood experiences and upbringing, social skills and emotion management skills. (Cacciatore, 2008; Sinkkonen, 2008). I work as a head instructor in a child caring institution for juveniles and also as a social worker in a comprehensive school and high school. Every day I meet children and teenagers with inadequate aggression managing skills. Some of them are violent towards others and/or themselves, some withdraw from social relations or suppress anger in other harmful ways. Most of my adolescent patients are under social services; some of them are under institutional care.

There is great potential in equine-assisted activities and therapies (EAAT), meaning both educational and therapeutic interactions with horses, especially when working with young people. There are specific elements in equine interactions that cannot be replicated artificially, and which seem to make them particularly effective. This literature review surveyed existing studies and literature about emotion and aggression education in adolescence and how EAAT is used in this area. The topic is approached from a psychosocial context.

AGGRESSION: BIOLOGICAL MECHANISM, WAY OF COPING AND SOCIAL CONTEXT

The study of aggression involves many theories and interpretations. In this review aggression is seen as a powerful and important force of living that can either be beneficial or harmful. This force is channeled for different purposes that can improve one's development or disturb it. Aggression is crucial from a biological perspective; it helps an individual to cope, to defend and to survive. Aggression can be reaction to frustration, disappointment or it can be a social message to other people. Healthy and beneficial aggression is a force that helps one to do something that needs power and encouragement. (Sinkkonen, 2008). Healthy aggression plays a very important role in puberty; it is the force driving the teenager to detach oneself from dependency and to become a young adult. Aggression can also help a person to state opinions, to defend her or himself and to have courage to live (Nurmi, 2013).

Anger is a powerful feeling and involves expressions such as anger, hate, and rage. Its effect on behavior depends on the capability to handle this emotion. If anger leads to rage, it can be a sign of inability to manage emotions (Nurmi & Schulman, 2013). Young people who have problems controlling aggressive feelings need knowledge, constructive restriction and support from adults. To be able to recognize and name different feelings is a very important skill. What cannot be understood cannot be fully controlled. Emotional skills are also connected to empathy and social skills. Young people who recognize more feelings are less likely to act aggressively, they are more accepted by their friends and they have better social skills (Cacciatore, 2008).

Aggression is also connected to fears. The greatest fear can be the fear of one's own blocked aggression that is projected onto external objects. It is like an inner "bogeyman" that is feared the most. The bogeyman is a prototype of self-developed projection of aggression (Yrjölä, 2015).

Aggression in Childhood and Adolescence

Emotional skills and skills related to aggression management develop in childhood. Constitutional traits, temperament and the relationship to the primary caregiver all contribute to emotional skill-building. The way parents help and support a child to come to terms with feelings of disappointment, anger and frustration is very important. Parental example, a stable and safe environment, and opportunity to practice all emotions are important factors. This helps the child to gradually learn to self-regulate (Cacciatore, 2008). Balanced growth requires long-term relationships with adults who understand and sufficiently support the child's development. Problems with aggression management skills can occur if the child has an exceptionally high need of support or if the care is insufficient and the demands of the environment are too high for the child's capability (Kaltiala-Heino, 2013b).

Adolescence is a vulnerable time for aggression challenges. In childhood, behavior control is mainly external, such as exerted by parents, but in adolescence the focus on controlling behavior gradually become the adolescent's own responsibility. Different kinds of disruptions in this process, for example, child neglect, mental disorders, developmental challenges and psychosocial challenges, may result in a level of self-control that does not correlate with normal age development. The environment and the adolescent's own demands and expectations may mean that behavior control fails (Kaltiala-Heino, 2013a). Adolescents who lack social skills, emotion management skills and have an impaired moral development level are more likely to engage in aggressive behavior patterns. In addition, if the environment supports violence it gives the wrong example. Aggressively behaving adolescents often think that violence is power and that it is synonymous with protecting one's own rights. When it is impossible to engage in problem solving and there is a lack of social skills, there is no other way to act than aggressively (Kaltiala-Heino, 2013c).

Aggressive behavior has increased among children and adolescents in the postmodern world. There is, of course, an increase in information, diagnostics and theories, but professionals and specialists working with children see the change (Sinkkonen, 2008). Sinkkonen (2008) suggests that our living environments have become more restless, troubled and stressing, and that children's central nervous systems become overloaded as a result.

A challenge to the development of emotional intelligence in a postmodern world might be caused by virtualization. Socializing has become something that is done from a distance. We can be very social in virtual exchanges, but fine-tuning of social skills and social intelligence happen in concrete interactions that are face-to-face. Adolescents are very social, but if social activity is narrowed and one-sided in a virtual world, there can be a lack of natural skills development. In typical face-to-face interaction the brain's "highway of emotional skills" controls behavior and ensures that it is socially accepted. In a virtual interaction this is missing so the neural networks do not get the feedback needed to develop adequate social skills (Goleman, 2010).

Anxiety, restlessness and attention deficits among children are often diagnosed to be neurological or somatic problems, but they can also be a sign of challenges to regulate emotions. In many cases there is also imprudent or aggressive behavior. The huge amount of information from the internet and social media can be hard for a child's brain to filter, self-regulate and to shield against. Some children can be very vulnerable to this kind of overload (Manninen, 2015).

Reactive Aggression and Proactive Aggression

Aggressive behavior in childhood and adolescence can be divided in reactive aggression and proactive aggression. Reactive aggression is short-term behavior where one feels that the environment is threatening and hostile. It is important to train emotional skills, learn to read the intentions of others, handle impulses, and have a meaningful, safe and confidential relationship with an adult (Soisalo, 2012). Proactive aggression is more complex and calculating. This means engaging in aggressive behavior on purpose, because it seems the right thing to do. Behavior may look like bossy, forcing and humiliating bullying towards others. With these adolescents it is important to reinforce positive behavior, to have systematic sanctions, to help them understand reason-consequence and to show alternative ways of coping (Soisalo, 2012).

Aggressive behavior in adolescence can also be a sign of a self-destructive way of living. In these cases individuals either withdraw from social interaction or behave violently and aggressively. Outwardly aggressive behavior is easy to notice but the real cause behind the violent reaction is not seen. Self-destructive ways can also be turned inwards and is then harder to recognize (Lämsä, Kiviniemi & Pönkkö, 2009). Aggressive behavior can also consist of passive behavior (passive aggression) such as manipulation, holding a grudge, not doing something on purpose, not taking responsibility, lingering, guilt tripping or being a martyr (Kaltiala-Heino, 2013c).

Behavior Disorders in Adolescence

Diagnosed behavior disorders have become more common during the last decades. There are many theories as to why this is, some of them neurobiological, some socio-cultural or socio-historical. Behavior disorders are connected to aggressive behavior. In addition, asociality, substance abuse, low income and low socio-economic status or asociality of parents can predict the child's future behavior disorder (Friis, Eirola & Mannonen, 2004; Lehto-Salo & Marttunen, 2006).

Disruptive behaviors are classified into different conduct disorder subcategories in Finland (Lehto-Salo & Marttunen, 2006): inter-family, asocial, social, oppositional defiant and depression-type conduct disorder. Oppositional defiant disorder is considered mildest of these categories. The more severe the disorder, the more likely there is aggressive behavior. Disruptive behaviors can also be related to neuropsychological problems. In adolescence, behavior disorders have some common features such as impaired expression of empathy, tendency to misinterpret others motives, and inability to feel guilt or remorse. Aggression and hyperactivity are also common features. Self-confidence is often poor and overcompensation common. The ability to encounter disappointments and control impulses is weak. Suicidal thoughts are more common than for the average adolescent.

There are some differences in the symptoms between males and females with behavior disorders. Young men often conduct criminal acts and are overall more violent. Young women tend to suffer mood disorders, anxiety, eating disorders and somatic disorders (Lehto-Salo & Marttunen, 2006). Behavior disorders in adolescence often predict psychiatric illnesses in adulthood.

Treatment of behavioral disorders is very challenging as these disorders are multifaceted. Treatment retention is low. It can also be very difficult to get the adolescent to trust an adult so that a traditional psychotherapeutic relationship can be formed (Lehto-Salo & Marttunen, 2006). Adolescents with behavior disorders also tend to partner and have children at an early age without a settled relationship (Lehto-Salo & Marttunen, 2006). Behavior disorders are thus often transferred to the next generation.

AGGRESSION EDUCATION AND EMOTIONAL SKILLS TRAINING

There is a natural need in humans to have an emotional balance. This means to regulate feelings in accordance to feedback from self and others. Emotion regulation is not an easy task. How well we know ourselves, the way we are, feel and act, and how we are in relationship to others is what makes a human a social creature. Emotional skills can help the development of "me" in the group of "us", and, the lack of these skills can make life heavy, lonely and unbearable.

Aggression is also connected to the experience of bad and evil. There can be an obsessive need to block out negative things and thoughts that can lead to fighting against oneself. Only by accepting both positive and negative sides of human nature can one live a full life. Through horses and horseback riding one is connected to both sides: one's good side where positive skills are known and one's bad side with things that are unknown or undesirable. Both sides are important and are part of life (Baum, 1994).

RECOGNIZING AND REGULATING EMOTIONS

Difficulties with emotion regulation can be divided into two categories: under-regulation and over-regulation (Kokkonen, 2010). Under-regulation may result in depression, anxiety, mania or can cause harmful behavioral patterns like aggressiveness. Over-regulation can result in emotional inhibition, which can cause difficulties in forming and maintaining meaningful relationships. In addition, it can lead to stress reactions, fears and obsessive-compulsive disorder (Kokkonen, 2010).

Adolescents with psychiatric symptoms often have problems recognizing and understanding emotions. Reacting to emotions, expressing or controlling them can be impaired. Physical appearance can reveal insufficient emotional skills. Physical tension, insufficient proprioception, bad posture and weak, unstructured quality of motion and movement are signs (Palola, 2013).

When an adolescent acts impulsively, uncontrolled and violently, it is often related to a harsh childhood involving experiences of abandonment. There can be severe cases of child neglect and traumas resulting from interpersonal interaction and attachment to the primary caregiver (Sinkkonen, 2008). Traumatic incidents leave a mark in brain development and become a negative model of how to feel and behave with other human beings. Sometimes there is a feeling of fear behind adolescent aggressive behavior. It can be a justifiable fear of abandonment, not fitting in, fear of controlling one's own life or fear of becoming an adult. In EAAT the atmosphere must be safe and open, so that the adolescent can face the fears and learn to notice, tolerate and control them with the help of the therapist/facilitator. When fears become tolerable, aggression is not needed to cover the fears (Tuomivaara, 2011).

Harmful experiences in life, especially violence, can cause chronic anger. Chronic anger is as common a problem as depression and anxiety, but is not treated even closely as much (Parkkinen & Tuomisto in Nurmi, 2013).

THE CHALLENGE OF INTERVENTIONS

It can be easy to identify aggressively behaving adolescents, but it is a challenge to change that behavior. Aggressive behavior is often self-reinforced enormously and feeds itself (Röning, 2013). Traditional "talk therapies" seem to be the most common but also the most ineffective intervention (Röning, 2013).

There are a few well-known interventions for aggressive behavior. The first one is Aggression Replacement Training (ART). ART was developed by Arnold Goldstein in the U.S. in the 1980's. This intervention is based on the idea that aggressive behavior has many different reasons, and aggression education should be a multi-faceted intervention. The main educational targets in ART are:

- practicing social skills, such as listening and participating conversation, and practicing to function in a group
- anger management, meaning finding socially approved ways to respond to provocation
- moral reasoning, when adolescents have to encounter different moral problems and reasoning levels, and work in a group to solve the moral problem

The second popular intervention is Teen Anger Management Education (TAME). This is also a group intervention that includes different relevant elements for anger management. These elements are:

- practicing interpersonal relationships instead of aggression
- identifying distorted feelings
- practicing problem solving
- practicing empathy
- learning to guide oneself through self-talk

Adolescents are taught to identify physical signs of aggression from the body and how to calm body and mind. In Finland, there is also third important aggression managing process when working with aggressive adolescents. This is called EVA and it is a psychiatric research and treatment unit in a hospital for aggressive minors. This intervention is used with teenagers with severe psychiatric problems and is used in a psychiatric ward. This intervention happens in a hospital environment, so that the conditions and the work space are highly safe and appropriate, risks are minimized, the risk of violent behavior is carefully analyzed, and the staff is trained to work with violent teenagers. The main elements are supporting calmness, intervening in an early state before agitation, physical intervention when needed and the processing of violent situations afterwards together with the staff and the adolescent (Röning, 2013).

Aggression Education and EAAT

There are many programs and studies that use equine-assisted methods with children and adolescents suffering psychosocial, emotional or behavioral challenges or learning disabilities. This being said, it can be quite difficult to gather and compare information of this field. The first challenge is to find information, especially in terms of determining whether study methods are scientifically objective, and how to know which sources are of good quality. The second challenge is similar to that of other "soft sciences": the prevalence of qualitative research of poor quality. All in all, there could be more international co-operation considering literature and studies in the field of EAAT.

There are very few studies of purely equine-assisted aggression education. Generally the main focus in the studies I found is on self-esteem, depression, learning disabilities or social exclusion and asocial behavior. These studies include elements also used in aggression education.

Ewing, MacDonald, Taylor and Bowers (2007) did a study on equine-facilitated learning for youth with severe emotional disorders. They used both qualitative and quantitative methods. The study was a nine-week program, two hours at a time, twice a week. The program was designed to teach skills such as cooperation, trust and responsibility. Each participant was paired with a horse and a volunteer. Quantitative results of this study indicated that there were no statistically significant positive changes in self-esteem, empathy, internal control, depression and feelings of loneliness. However qualitative measurements; interviews and observations by teachers, instructor and volunteers indicated remarkable positive changes after participation.

Schultz, Remick-Barlow, and Robbins (2006) studied equine-assisted psychotherapy addressing self-esteem and personal confidence, communication and interpersonal effectiveness, trust, boundaries and limit-setting, and group cohesion with children. The study was done over an 18-month period. All participants showed improvement in GAF (Global Assessment of Functioning) scores and a statistically significant correlation was seen between the percentage improvement in the GAF scores and the number of sessions given. The greatest improvements were with the youngest participants and those who had a history of physical abuse and neglect. Equine-assisted psychotherapy appeared to be effective for children with adjustment disorder, mood disorder, post traumatic stress disorder, ADHD, and disruptive disorder.

Kaiser, Smith, Helenski, and Spence (2004) did an observational 8-sessioned therapeutic riding program with 17 at-risk children in special education programs. These children were considered at-risk for poor performance in school or life. The measurement of the results were psychosocial, measuring anger, anxiety, self-competence, physical coordination and, with special education children, anger and cheerfulness. Boys in the special education program had decreased anger after the program and their mothers' perception of the boys' behavior was that it was positively affected.

Bronkhorst (2006) did a master thesis on horse whispering for the aggressive child. It was a literature review on how animal assisted therapy and equine facilitated psychotherapy is used for treating inappropriate, aggressive behavior by a young boy. Horse whispering was based on the idea of nonverbal communication and learning to understand and interpret the horse's communication and to communicate with the horse. This research suggested that horse whispering can be a potential therapeutic element when breaking the aggressive behavior cycle by helping promote non-aggressive ways to assert oneself and to form attachment and relationship.

Bachi, Terkel and Teichman (2012) studied equine-facilitated psychotherapy for at-risk adolescents. The main target was to study self-image, self-control, trust and general life satisfaction within two groups, the treatment group and a control group which did not receive equine-facilitated psychotherapy. The study found that there occur positive changes in all studied areas. Even though the results in means and measurement were not significantly different, there was a clear trend of increase in trust, self-control, self-image and general life-satisfaction. The change was noticeable in the treatment group.

Why Equine Therapy: Effective Elements

The most challenging part of this study was to find the connection between equine-assisted work and aggression education theories. What are the elements that can potentially create change and why use horses with clients who have aggression management difficulties? Simply said, there are multiple target areas that are affected simultaneously.

Physical Sensations and Body Awareness

EAAT can create positive physical sensations and generate body awareness. Moving increases body temperature and affects muscle relaxation, which can enhance one's feelings of well-being. In EAAT the horse's stimulation, rhythm, warmth and movement, give sensory impulses and create tangible body activation. By helping clients become aware of this, increased body awareness can happen. This helps strengthen body control, self-image and feeling of life control (Tuomivaara, 2011).

Touching is one important therapeutic element that is used with children and teenagers who are receiving treatment as part of being in child protection in Finland. But who can touch, how and when, is very important when using it as a therapeutic method. With horses touch is less complicated, because the horse is neutral or positive to have physical contact with.

Synchronization

Being with a horse and finding a synchrony of being and moving is important. Social psychologists have discovered (Goleman, 2009) that the more both members of the party of two naturally synchronize their movements, the more positive emotions this encounter generates. Non-verbal elements and physical synchrony maintain the harmony, even though there is no talking or there are interruptions. When two persons have this positive flow of signals it amplifies their resonance. They give and receive moods. Synchrony gives a feeling of mutual understanding. Physical synchrony gives a sense of primitive pleasure (Goleman, 2009). This happens by moving in rhythm. Even babies adapt their movements in the womb to the rhythm of mother's voice. Newborn babies also have this kind of resonance with their mother. So having a synchronized movement and rhythm with a horse can provide quite primitive feelings of satisfaction. The horse's physical movement can also help getting in touch with inner movement. Humans' external and internal movements are related, and a physical impulse can generate an inner impulse. Emotions and physical movements are interrelated (Mehlem, 1994).

Motivation

One very important, one could say essential, factor in treatment for adolescents is motivation. Motivation is not something that a therapist can give or make happen; it is an internal state of will.

THE POTENTIAL OF THE HORSE IN ADOLESCENT AGGRESSION MANAGEMENT: A REVIEW OF THE LITERATURE

Adolescents with challenging backgrounds can have a very negative perspective on life and other people. Motivating them for treatment is difficult. This is where the horse makes a difference. Horses can be feared, but they are also seen as powerful and fascinating creatures. It can also be easier for the adolescent to connect to something other than another human. The horse has a sensitive, almost human like nature, large size, soft eyes and a sociocultural, historical status like no other animal which gives it potential for interaction. There are many reasons that can help explain why a horse can produce motivation even within those adolescents who cannot or will not commit to any other therapy.

EMPATHY TRAINING

EAAT offers empathy training. Empathy cannot be learned by talking, it comes from experience and social knowledge. Children with child welfare backgrounds may have had very damaging experiences that can cause weakened social skills. Attaching to an animal can be a first significant sign of recovery. It is safe to show affection toward an animal, feel trust and experience empathy (Julius, Beetz, Kotrschal, Turner, & Uvnas-Moberg, 2013). To recreate empathy and to strengthen self-esteem constitutes positive feedback on one's significance to others. When a child feels important and meaningful, the desire to interact with others increases (Sinkkonen, 2013).

OVERCOMING CHALLENGES

Being with and working with a horse forces the adolescent to be confronted with frustration and disappointment. Destructive aggression arises when things do not happen as planned. This is where the horse and the therapist/facilitator can help the teenager feel disappointment and overcome it. Destructive or blocked aggression can cause feelings of guilt, evilness and underestimation. With a horse one must stop and think about the cause of disappointment and aggression. When there is time and opportunity to correct inappropriate and destructive behavior, one learns to take responsibility of own feelings and actions (Yrjölä, 2011). So it is not all about the positive experiences with the horse, it is essential to go through negative thinking and behavioral patterns so that they can be changed. When working with aggressive adolescents, the best "area" to work in is when the aggression has increased by some amount. If there are no aggressive feelings, the adolescent's performance may be good but no significant change can be produced. If the aggression has crossed the limit of tolerance and endurance, the situation may not be educational anymore but simply about surviving. This is where the adolescent cannot control behavior. The best working scenario is when there is enough emotional challenge that the adolescent can handle the burden and learns from the situation with professional help.

Potential Space

Being with a horse can open a potential space. Potential space, as proposed by Donald (Winnicott, 1987), means a loving atmosphere similar to that of a mother and her newborn. That atmosphere includes trust and safety, positive expectation, creativity, hopefulness, vitality and love. These positive elements are slowly adopted to be perceived within oneself. The potential space can continue through life and be present in different situations (Yrjölä, 2011). The socio-educational power is to open a potential space for the adolescent to form a relationship with the horse where she or he can encounter emotions in a safe, appropriate atmosphere.

Positive Experiences and Connectedness

At-risk and aggressive adolescents may not easily trust other humans. They can have so many disappointing experiences of interpersonal relationships that they only trust themselves and want to be able to live without any dependence on another human being. The aim of equine-assisted work is to form an interaction with the horse where the adolescent can get positive experiences of relating to another being and having a connection to someone. A goal is also to transfer horse-human relations into inter-personal relations. That way the horse can offer a feeling of connectedness (Ewing et al., 2007).

The horse functions as a companion and a co-therapist, so choosing the right horse is very important. The horse must be curious and willing to interact, but also needs to have boundaries when it comes to unacceptable behavior. The horse's temperament must be suitable for the client. This must be evaluated every time depending on the individual client. The main focus is on the horse's ability to offer remedial experiences.

Precautions

In some cases equine-assisted work cannot be recommended as an aggression management intervention. If an adolescent cannot be calmed in an aggressive situation without heavy physical restraining, other interventions must be considered first. The number one priority is that the safety of the patient, the horse and the therapist can be maintained. If the patient is too impulsive and there is a real risk for severely violent behavior, other more suitable interventions should be considered. In addition, if the patient is diagnosed as psychopathic and is unable to connect with an animal and have feelings of empathy there might not be any motivation to connect to the horse.

CONCLUSION

As a closing remark I return to social pedagogy. Martin Buber sees a human growing and evolving through "Me-You" relationships (e.g. Buber, 1993). There are also "Me-It" relationships that are more objective and intentional, but "Me-You" is a naturally primitive, basic and subjective relation that is essential for personal and social growth. Without "You" there is no "Me". The way I see the potential for EAAT when working on anger management in adolescents, is that the horse can be that significant "You" when there is no room for a human "You" at the time. The horse can be the first contact for the adolescent to want to become "Me".

REFERENCES

Bachi, K., Terkel, J. & Teichman, M. (2012). Equine-facilitated psychotherapy for at-risk adolescents: The influence on self-image, self-control and trust. Clinical Child Psychology and Psychiatry, 17(2), 298-312.

Baum, D. (1990). Psychisch kranke: Menschen auf dem pferd In M. Gäng (ed.), Heilpädagogisches Reiten und Voltigieren (pp. 236-240). München: Reinhardt Verlag.

Bronkhorst, K. (2006). The psychotherapeutic worth of horse whispering for the aggressive child. University of South Africa. Master's Thesis.

Buber, M. (1993). Minä ja sinä (trans. Pietilä, J.). WSOY: Helsinki.

Cacciatore, R. (2008). Aggression portaat. Opetushallituksen opetusmateriaali kouluille. Vammalan Kirjapaino Oy: Sastamala.

Ewing, C. A., MacDonald, P. M., Taylor, M. & Bowers, M. J. (2007). Equine-facilitated learning for youths with severe emotional disorders: A quantitative and qualitative study. Child and Youth Care Forum, 36, 59-71. DOI: 10.1007/s10566-006-9031-x

Friis, L. & Eirola, R. & Mannonen, M. 2004. Lasten ja nuorten mielenterveystyö. WSOY: Helsinki.

Goleman, D. (2009). Sosiaalinen Äly. (trans. Jänisniemi, L. & Paajanen, A.). Otavan Kirjapaino Oy: Keuruu.

Julius, H., Beetz, A., Kotrschal, K., Turner, D., & Uvnas-Moberg, K. (2013). Attachment to pets: An integrative view of human-animal relationship with implications for therapeutic practice, Cambridge, MA: Hogrefe Publishing.

Kaiser, L., Smith, K., Heleski, C. & Spence, L. (2006). Effects of a therapeutic riding program on at-risk and special educational children. Journal of American Medical Association, 228(1), 46-52.

Kaltiala-Heino, R. (2013a). Väkivaltakäyttäytyminen. In P. Nurmi (ed.). Lapsen ja nuoren viha. (pp.71-107). PS-Kustannus: Jyväskylä.

Kaltiala-Heino, R. (2013b). Aggressio lapsuudessa ja nuoruudessa. In P. Nurmi (ed.). Lapsen ja nuoren viha (pp. 47-68). PS-Kustannus: Jyväskylä.

Kaltiala-Heino, R. (2013c). Aikuinen mallina ja vihanhallinnan opettajana. In P. Nurmi (ed.). Lapsen ja nuoren viha (pp. 149-162). PS-Kustannus: Jyväskylä

Kokkonen, M. (2010). Ihastuttavat, vihastuttavat tunteet. PS-kustannus: Jyväskylä.

Lehto-Salo, P. & Marttunen, M. (2006). Käytöshäiriöt ja niiden hoito. In E. Laukkarinen, M. Marttunen , S. Miettinen, & M. Pietikäinen, (ed.). Nuoren psyykkisten ongelmien kohtaaminen (pp. 99–111). Duodecim. Karisto Oy: Hämeenlinna.

Lämsä, A-L, Kiviniemi, L. & Pönkkö, M-L. (2009). Kiusatun leima ja itsetuhoinen elämäntapa. In A-L Lämsä. (ed.). Mun on paha olla. Näkökulmia lasten ja nuorten psyykkiseen hyvinvointiin (pp. 97- 110.). PS-kustannus: Jyväskylä.

Manninen, H. 2015. Miksi levottomia lapsia on niin paljon? Lääkärilehti 43/2015 (23.10.)

Mehlem, M. (1994). Heilepädagogisches reiten als bestandteil der psychotherapie. In E. Törmälehto, E. (ed.), (trans. Virtanen). Ratsastusterapia: Kirjallisuusotteita toinen osa (pp. 77-81). Original title:Therapeutisches Reiten.

Nurmi, P. (2013). Lupa tunteisiin. In P. Nurmi (ed). Lapsen ja nuoren viha (pp. 15-28). PS-kustannus: Jyväskylä.

Nurmi, P. & Schulman, M. (2013). Aikuinen lapsen vihan peilinä. In P. Nurmi (ed). Lapsen ja nuoren viha (pp. 29-46). PS-kustannus: Jyväskylä.

Palola, L. (2013). Ratsastusterapian mahdollisuudet lasten ja nuorten psykososiaalisessa kuntoutuksessa. In K. Ikäheimo (ed.). Karvaterapiaa – Eläinavusteinen työskentely Suomessa (pp. 157-170). Solution Models House: Yliopistopaino Helsinki.

Parkkinen, L. & Tuomisto, M. (2011). Viha, vihaisuus, hyvinvointi ja terveys: Vihaisuuden merkitys, arviointi, diagnostiikka ja hoito. Suomen käyttäytymistieteellinen tutkimuslaitos: Tampere.

Schultz, P. N., Remick-Barlow, G. & Robbins, L. (2007). Equine-assisted psychotherapy: a mental health promotion/intervention modality for children who have experienced intra-family violence. Health & Social Care in the Community, 15(3), 265-271. DOI:10.1111/j.1365-2524.2006.00684.x.

Sinkkonen, J. (2008). Mitä lapsi tarvitsee hyvään kasvuun. WS Bookwell OY: Juva.

Sinkkonen, J. (2013). Lapsen yksilökehitys ja suhde eläimiin. In K. Ikäheimo (ed.). Karvaterapiaa – Eläinavusteinen työskentely Suomessa (pp. 37-43). Solution Models House: Yliopistopaino Helsinki.

Soisalo, R, (2012). Särkyvä mieli: Lasten ja nuorten psyykkinen oireilu. AS Printon Trukikoda: Tallinn.

Tuomivaara, P. (2010). Käytöshäiriöinen lapsi ratsastusterapiassa. In S. Mattila-Rautiainen (ed.). Ratsastusterapia (pp. 347-372). PS-kustannus: Jyväskylä.

Röning, T. (2013). Aggressiivisesti käyttäytyvien nuorten hoitomenetelmiä. In P. Nurmi (ed). Lapsen ja nuoren viha (pp. 211-220). PS-kustannus: Jyväskylä.

Yrjölä, M-L. 2011. Hevosen vaikutukset ja ihmiselle asettamat haasteet. In S. Mattila-Rautiainen (ed.). Ratsastusterapia (pp. 88-109). PS-kustannus: Jyväskylä.

Yrjölä, M-L. E-mail conversation 18.11.2015.

ARTICLE FOUR

Role of Learning Theory in Training and Handling the Therapy Horse

Elyssa B. Doner, M.A, IdyleWild Farm
Nina Ekholm Fry, MSSc., Institute of Human-Animal Connection
University of Denver

Correspondence to:
Nina Ekholm Fry, Director of Equine Programs, Adjunct Professor
Institute for Human-Animal Connection, University of Denver
2148 South High Street, Denver, CO 80208, USA
Email: nina.ekholm-fry@du.edu

ABSTRACT

Practitioners who provide services with horses that involve human-horse interactions operate in a complex environment. When horses are specifically included to promote human mental health and wellbeing, it seems particularly important that the practitioner has extensive knowledge of equine ethology, including how horses learn and the principles of ethologically sound training. Therapy horses interact with clients whose main purpose is not to learn specific handling practices but to gain personal insight through their relationship and interactions with the horse. When designing sessions or overseeing client activities, practitioners need to take care to not intentionally create distressing situations for the horse. Poor preparation for work role, lack of consistent conditioning or lack of appropriate training can inadvertently create high-risk, high-stress situations for clients and horses. We propose that knowledge of learning theory, which explains how horses learn through common learning processes such as habituation and sensitization, and classical and operant (instrumental) conditioning, is central to ethical interaction with the therapy horse, effective client services, and appropriate risk management for all involved in the session.

KEYWORDS

Therapy Horse, Equine Therapy, Learning Theory, Horse Training, Work Role, Equine-Assisted Practitioner

INTRODUCTION

While distinct in their approaches and goals, the areas that comprise the growing field of equine-assisted activities and therapies (EAAT) are founded on the notion that therapeutic and educative benefits can be derived from intentional interactions with this animal. The horse's kinesthetic and aesthetic features may supersede attributes of other species employed in animal-assisted activities and therapies in terms of physical versatility. However, it is the horse's cognitive abilities, learning capacity, and psychological and social characteristics that make equine-based psychological treatment strategies, here called equine-assisted mental health (EAMH) and equine-based learning and educational approaches, here called equine-assisted learning (EAL), especially impactful.

Horses have many cognitive attributes, learning mechanisms, and psychological and emotional dispositions similar to those of humans (Hahn, 2014; Wathan, Burrows, Waller, & McComb, 2015). When such commonalities are explained to an EAMH/EAL client, it can help her or him relate to a therapy horse as a fellow sentient being. For example, many of the ethological challenges faced by horses are similar to those of humans. Environmental challenges for the domestic horse, such as leaving the home range, encountering new stimuli (neophobia), and deviating from a known path to cross rather than avoid obstacles (McGreevy & McLean, 2010), can be viewed as equivalent, metaphorically or literally, to challenges confronted by a client, such as moving, starting a new job or new school, or overcoming obstacles to success in a particular environment. Similarly, social challenges for horses, leaving the social group, being asked to take the lead by its rider in the company of established leaders, or being forced to stay at the rear of the group (McGreevy & McLean, 2010), parallel social challenges encountered by their human counterparts. When clients are able to identify with their therapy horse on a deeper, more meaningful level, benefits in areas of relational skill building and development of healthy social behaviors can be gained.

Practitioners providing EAMH and EAL services are not only responsible for promoting the wellbeing of clients but also hold ethical responsibilities to their therapy horses. An understanding of equine behavior, mental processes, learning and overall welfare is necessary to provide ethical and effective services, which include the reduction of risk in the client session (Ekholm Fry, 2013).

Working with horses successfully and ethically requires a well-developed understanding of equine ethology, learning theory, and the goals and limitations of the training approach being used (Goodwin, McGreevy, Waran & McLean, 2009). Ethology, or the study of animal behavior in a natural environment, provides insight into equine behavioral needs and preferences, value systems, and motivation (McGreevy & McLean, 2007). Cognitive ethology is a relatively recent emergence in animal-learning studies, defined as a "multi-disciplinary blend of behaviorism, cognitive psychology, neuropsychology, cognition, and ethology" (McGreevy & McLean, 2010, p. 57). Understanding of cognitive ethology is essential for the competent horse practitioner, especially one who facilitates human-horse interactions such as EAMH or EAL.

In this article we focus on the role of learning theory in training and handling the therapy horse. Learning theory explains how horses learn through common learning processes such as habituation and sensitization, and classical and operant (instrumental) conditioning. As such, the purpose of this article is not to provide a complete overview of these concepts but to introduce them as useful for the equine-assisted practitioner.

Definitions

We use the umbrella term equine-assisted activities and therapies (EAAT) to include physical, occupational, and speech-language therapy conducted with horses, as well as learning-based approaches and psychotherapy that include horses. We use the term equine-assisted mental health (EAMH) to specifically mean psychotherapeutic treatment strategies that involve horses. These are provided by mental health professionals who are credentialed to provide psychotherapy or counseling services. We use the term equine-assisted learning (EAL) to mean services involving horses that promote learning, education, development and growth, which may or may not include riding or specific horsemanship skills. The term practitioner is used to mean a person providing equine-assisted services, whether these are EAMH or EAL. The term client is used to span both the traditional use of patient in treatment, and the terms student and participant in educational services. The term therapy horse is used to describe equines involved in both therapy and educational services for the convenience of this article.

Overview of Learning Theory

Training animals is a complex task. It involves gathering knowledge about a species different from our own and the capacity to apply this knowledge in a situational, individualized, and practical way. Training animals also requires the ability to manage one's own state of mind so that humans and horses involved in the training situation remain in a state conducive to learning.

In scientific terms, the process of training can be defined as the "intentional modification of the frequency and/or intensity of specific behavioral responses" (Goodwin, et. al., 2009, p. 10) in the horse. Such modifications are attained by the shaping of a desired response through non-associative learning, such as habituation or sensitization, or associative learning, such as reinforcement or punishment (Murphy & Arkins, 2007). In the latter process, specific behaviors are reinforced in a step-by-step fashion toward a specific end point, and undesirable responses are suppressed and extinguished using punishment (Goodwin, et. al., 2009). For EAMH and EAL practitioners who are working with human health and development, it may seem foreign to imagine the use of punishment. However, in the scientific language used in learning theory, the term punishment is void of emotional connotation and is simply used to describe a process that is likely to decrease the frequency of behavior. For instance, ignoring an unwanted behavior, which is considered a form of punishment in learning theory, is commonly used in child rearing.

While horse training may be complex, and many methods are promoted, its principles can be accessed and learned. This is especially important for the EAMH/EAL practitioner since the main objective in typical horse training methods, which are generally goal-driven, is to produce a safe and effective animal for a particular task. Moreover, horse handling and training cannot fully be separated as the animal is constantly learning, even in interactions where the human does not consider herself actively training the horse. For example, when the horse is brought from the pasture to the arena, and when the client interacts with the horse during a session, the horse is learning and behavior is shaped. Through understanding the basic components of learning theory, one can become an effective and ethical trainer, and adept at assessing the impact that horse-handling and training may have as clients work with therapy horses. Such knowledge not only supports a relational view of the world, which EAMH and EAL practitioners typically hold, but also incorporates the concepts of relationship-building and clear communication.

Habituation and Sensitization

Habituation and sensitization create or extinguish behavioral responses based on a single stimulus and are classed as non-associative learning. Habituation, which includes desensitization, is a basic and important process during which the horse learns what elements in its environment are familiar and safe, and, thus, do not warrant significant action or response, versus elements that are considered dangerous or potentially dangerous and produce fear-driven reactions. These environmental elements include the array of equipment used in human-horse interactions such as saddles, horse boots, cones and barriers, including electrical fences, as well as the presence of a human on its back (McGreevy & McLean, 2007).

> Habituation is a learning process whereby, after repeated exposure at the same intensity to... (a) stimulus, an animal becomes accustomed, and its reaction to that stimulus either diminishes or disappears. A stimulus may be anything that incites a response. Although this is an extremely simple form of learning, it is, nonetheless, extremely important because it allows an animal to subconsciously filter out non-vital information in its environment and focus instead on what is significant. (Hanggi, 2005, p. 247)

Sensitization involves the same learning process but results in the opposite effect. Sensitization creates significance and a quick response to something in the environment. The objective when sensitizing a horse is to create a desired response with minimal application of a discriminative stimulus, such as directive pressure with a lead line. The process of sensitization involves increasing the magnitude of a discriminative stimulus or cue until a particular response occurs. At that point, the stimulus is immediately removed. This process is often facilitated through the use of negative reinforcement (see section below), such as when teaching a horse the cue to stop:

> Sensitization to a light stop signal demands the same process as the initial [negatively reinforced] training of the stop signal. It involves increasing the intensity of the discriminative stimulus (the light stop signal) until a sufficient response occurs, in which case the stimulus is immediately withdrawn. (McGreevy & McLean, 2010, p. 65)

This procedure is repeated until the horse responds directly (is sensitized to) minimal pressure on the reins when asked to transition to a halt.

Effective use of habituation processes reduces stress in the therapy horse. It helps the horse quickly get familiarized with equipment and other standard features of the work environment so that the horse does not have to expend energy on preparing for or engaging in flight or fear behaviors. Sensitization helps the therapy horse pay attention to things relevant to its work role so that it can respond when given the slightest cue and not experience increased pressure or other outcomes. See McLean (2012) for a brief overview of habituation methods.

Operant and Classical Conditioning

Associative learning occurs when a relationship between at least two stimuli has been established, and these stimuli have occurred close to each other in space or time. Operant and classical conditioning are associative learning processes. A horse starting to canter from a leg or vocal cue, or walking next to a human without being attached to a lead rope are both the result of associative learning.

Operant conditioning focuses on the ways in which the animal operates in its environment. It consists of presenting or omitting reward or punishment when an animal makes a specific response to a particular stimulus. As such, it enables an animal to associate events over which it has control (McGreevy, 2007). Operant conditioning enables an animal to make changes to its environment (that is, receiving a reward or avoiding punishment) by voluntarily responding to a specific stimulus. A horse may learn to open a gate that leads to hay storage (receiving reward) or move away when another horse is approaching (avoiding punishment).

An early example of classical conditioning is colloquially referred to as "Pavlov's dogs". Dogs started salivating upon merely hearing a sound that had been introduced just before receiving food on a regular basis (Pavlov, 1927), thus salivation had become paired with the sound. Classical conditioning creates an association between an unconditioned stimulus, which is something irrelevant (a sound in case of the dogs), and something relevant (being fed), producing a conditioned stimulus (the sound) and a conditioned response (salivation). The process of classical conditioning makes it possible for horses to learn the meaning of certain vocal or gestural cues (such as whistling or pointing) and perform actions based on what the discriminative stimuli have come to mean.

The fundamental difference between operant and classical conditioning is that classical conditioning enables the animal to associate events over which it has no control. It establishes a response to a new stimulus by repeatedly pairing it with an already established stimulus that evokes innate or learned psychological or behavioral responses (McGreevy & McLean, 2010). In operant conditioning, rewards become associated with responses, and responses are deliberate and voluntary. In classical conditioning, rewards become associated with stimuli, and responses to these stimuli are automatic, and, thus, involuntary (McGreevy, 2007).

Punishment and Reward

During any kind of conditioning or behavior modification, an animal provides a response in order to obtain a reward, known as reinforcement, or avoids giving a response to preclude punishment. The primary purpose of punishment is to suppress an undesirable behavior. Punishment can be either positive (something aversive added) or negative (something attractive withheld), and is generally perceived as the least popular approach to behavioral modification as it does not on its own provide direction in terms of behavior that will be rewarded. Although its definition – the presentation of a stimulus to suppress a behavior – is objective in nature, punishment is often associated with certain subjective, emotional states including fear, frustration, and pain. Depending on how punishment is used, examples range from simply withholding a food award until the horse has performed the desired behavior (in positive reinforcement, see below), to inflicting severe pain in the horse, punishment can comprise an array of problems. Punishment alone is:

> suppressing but not enhancing alternative behaviors; has the potential to desensitize an animal to the punishing stimulus if the punishment intensity is not optimal; carries the risk

> of deleterious emotional changes that can interfere with attention and learning; and may be associated by the animal with the person delivering it. (McGreevy & McLean, 2009, p. 194)

Punishment as a means of behavior modification should be used with caution as it has the potential to interfere with the learning state and welfare of the horse.

The primary function of reinforcement is to promote or strengthen a desired response. Reinforcement can be either positive (receiving something desired) or negative (avoiding something unpleasant that reduces comfort). Both types of reinforcement serve to "strengthen the relationship between a stimulus and the desired response so that when the same stimulus is given again, there is an increased chance that the animal will repeat the response" (Hanggi, 2005, p. 248). Punishment and reinforcement techniques both require maintenance in order for a response to become firmly established and eventually habitual (Miller, 1996), thus allowing humans and horses to perform complex tasks together.

A Further Exploration of Negative and Positive Reinforcement

The rewards or reinforcers that are used in both negative and positive reinforcement approaches can either be primary or secondary. As McCall and Burgin (2002) explain, "primary reinforcers reduce physiological needs (e.g., hunger, thirst or escape from aversive stimuli) that are common to all members of a species" (p. 254). In negative reinforcement, the primary reinforcer has to do with maintaining comfort. Negative reinforcement is the primary means of shaping equine behaviors (McCall & Burgin, 2002). Typical primary reinforcers in positive reinforcement are food and scratching. Primary reinforcers tend to be inherently understood so that the animal needs no training to comprehend the reinforcement. As such, they are the fundamental components of operant conditioning. Different primary reinforcers hold different motivational value for different individuals at different times. The difference between positive and negative reinforcement has to do with selection of a primary reinforcer.

Secondary reinforcers do not reduce physiological needs and often are referred to as conditioned or learned reinforcers. They include any vocal or gestural cues that do not hold inherent meaning to the horse (whistles, clicks, words, body positions, et cetera).

> Secondary reinforcers acquire their reinforcing properties from their association with primary reinforcers through Pavlovian (classical) conditioning, and the effectiveness of the secondary reinforcer may be directly related to the amount or value of the primary reinforcer with which it was paired. (McCall & Burgin, 2002, p. 254)

McCall and Burgin (2002) conclude that, "from a practical training standpoint... secondary reinforcers are useful in horse training situations, but they should be paired periodically with a primary reinforcer to maintain their effectiveness" (p. 261).

Negative Reinforcement (NR). A fundamental behavioral need for horses is comfort (McGreevy, 2007). Based on this understanding, horses typically are trained to perform actions by providing them with a means to avoid something unpleasant (Hanggi, 2005). For example, McGreevy and McLean (2010) describe the use of NR in foundation equitation training. Here, the trainer seeks to establish control over the horse's basic locomotory responses (including go forward, turn, go sideways, and stop/slow/go backward), while the horse strives to meet his behavioral need for comfort. An effective trainer establishes this control via NR, commonly known as 'pressure-release', meaning that the release of the pressure has to come at the exact time the horse starts to perform the desired behavior in order to maintain the agreement between trainer and horse with regards to behavioral outcomes. In other words, to achieve control of the basic locomotory responses via NR, the trainer identifies a targeted response. Performance of the targeted behavior results in an immediate removal of the pressure/discomfort, which had been applied to initiate behavior (i.e., the cue).

It is this reliance on pressure and the release of pressure that distinguishes horse training, especially for riding, from training of most other species (McGreevy & McLean, 2010). With the exception of horse trainers, most animal trainers rely primarily on positive reinforcement (PR). Animals including dolphins, seals, bears, large cats, and dogs are, to a considerable extent, trained via PR. One reason for this is because the primary reinforcer can be provided at a distance, and the "pressure-release" component of NR is not possible to use in such circumstances. Under-saddle training, for example, cannot be performed at a distance, so PR is not a viable training option (see section below for discussion of using PR while riding). Additionally, when using PR, the trainer has the option of ignoring an error made by the animal. If the animal moves away, for example, the trainer can wait until the animal chooses to return to the lesson. Ignoring such errors is not ideal when working with horses, particularly when training under saddle. The horse's size and work role often precludes simply waiting for it to return to a calm state, and a very excited horse creates dangerous situations whether the trainer is on the ground or astride. These examples delineate why horse training is historically contingent on proper application of NR – the trainer does not have to wait for the horse to offer the response – she can motivate the animal to effectively trial a response (McGreevy & McLean, 2010). In horse training, NR forms the foundation for control and consistency. Horses excel at this type of learning, especially when PR is also used (see section below).

The main issue with negative reinforcement is also part of its effectiveness: use of pressure. Constant, long-term pressure through incorrect use of negative reinforcement has severe health implications for the horse and can lead to long-term stress. In other words, pressure is a tool that one must receive training in using correctly (Ekholm Fry, 2014), and incorrect application of negative reinforcement has a larger, unfavorable impact on the horse than incorrect application of positive reinforcement.

Positive Reinforcement (PR). Positive reinforcement can be defined as "the addition of a pleasant stimulus (a reinforcer) to reward the desired response and thus make this response more likely in the future" (McGreevy & McLean, 2010, p. 283). While it is far less commonplace than NR techniques, PR is used in training horses and understanding its principles will create effective overall training (Hanggi, 2005). Innes and McBride (2008) studied of the effects of NR and PR training protocols on rehabilitating horses. They found that NR, as used in their study (unpleasant stimulation through tapping with a riding whip) did not "overtly compromise the [horse's] welfare" (p. 367) and the use of PR training (clicker training paired with food as the primary reinforcer) "confer active benefits to the animal" (p. 367). Further, horses that participated in the PR regime "tended to be more motivated to partake in [training] activities and, as a result of performing more explorative/'trial-and-error' type behaviors" (p. 367). This indicates that PR can help shape desired behaviors in horses.

Sankey, et al., (2010) tested the effects of PR on the ability of 23 young horses to recall an association with a specific partner, and learn a new task. Trials (using food pellets as a reinforcer) generated beneficial reactions towards humans during training as well as several months later. Contact and interest with specific and novel humans both during and after training were noted, and PR was deemed to enhance learning and memorization of the task itself.

Although these and other current research indicates that PR is an effective method to apply in equine behavior modification, a number of issues exist with PR as it pertains to horse training. McGreevy and McLean (2010) point out that PR "generally relies on food as the reinforcer, and food is particularly difficult to deliver immediately after the ridden horse offers the desired response" (p. 77). McLean (2004) found that brief time delays (of 10 seconds) between the horse providing a desired response and the subsequent reward appear to cause a significant degree of difficulty for the horse. Reward devices that deliver food via a remotely controlled hollowed bit have been used successfully in experimental situations; however, such devices are rarely if ever seen outside of scientific studies.

While proponents of PR techniques including E.B. Hanggi are calling for "more research and public demonstrations... to educate horse handlers on the techniques and efficacy of [PR] methods" (Hanggi, 2005, p. 249), Murphy and Arkins (2007) observe that "curiously, within the disciplines of equitation, learning of the desired behavioral processes is most often attempted using NR strategies... On the other hand, experimental designs to actually measure, assess, or induce learning and desired responses in the horse rely almost exclusively on primary PR regimes" (p. 10). This observation embodies serious implications relative to the efficacy of PR and "real world" training techniques. If training for research purposes is predominantly based on PR, the results of the trials can be considered only tangentially relevant to everyday training. Scientific research depends on controlled environments that are not normally available for common training regimes. With this in mind, it is not surprising that PR techniques are often less than conclusive when applied outside of scientific trials and have yet to be recognized as more mainstream training techniques for horses.

A final observation about the use of PR in horse training regimes relates to equine ethology. As indicated earlier, the vast majority of animals including dolphins, seals, bears, large cats, and dogs are, to a considerable extent, trained via PR and are all predators. As such, they are genetically driven to focus their attention on a target – prey, or more simply stated, food – and move directly towards it to attain it. For these animals, the use of PR makes perfect sense as a training technique since it speaks to the natural drive of the predator. Horses, as prey animals, are not genetically driven to focus on a specific target as food is procured and consumed continuously. Rather, they have an expansive awareness of their surroundings and are driven to survive primarily through movement away from danger (flight). In this sense, the use of PR does not appear to be directly aligned with equine ethology. Motivation is also important to consider from a species-specific point of view, both in terms of effectiveness in the process of trialing behaviors (depending on how and where NR pressure is created it gives the horse a clue as to what the desired behavior might be, and less trialing, which expends energy, is needed), and in order to override a behavioral conflict in as to execute the desired behavior. Murphy and Arkins (2007) note that "food reinforcement is widely used as a positive stimulus in equine learning and behavioral experimental trials, although freedom from aversive stimuli has been judged as more reinforcing than food provision in the horse" (p. 6).

Strategic Use of Both Reinforcement Approaches. Ideally, practitioners should incorporate informed and strategic use of both NR and PR into a well-balanced training program (Hanggi, 2005, p. 249) as they both have advantages and can effectively be used together. McGreevy and McLean (2010) note that clicker training, a PR method, is generally used to help maintain the acquisition of a light signal after establishing desirable, consistent responses through NR. This also implies that PR is often predicated on effective applications of NR.

It seems that the most salient scientific trials and training regimes are those that are designed to examine and promote equine learning under conditions reflecting equine adaptive behavior. In this sense, equine ethology is the key to finding more effective and illuminating approaches for research and training programs in the future (Goodwin, et. al, 2009).

Learning Theory and the EAMH/ EAL Practitioner

We wish to highlight three areas where knowledge and application of learning theory plays an important role for the EAMH/EAL practitioner:

1) in designing and overseeing activities in sessions;
2) in maintaining mental and physical health of the therapy horse; and
3) in preparing the therapy horse through establishing work role.

Designing and Overseeing Activities in Client Sessions

The therapy horse is typically seen as a source of support and relationship, as well as a facilitator of insight and change in the client session (Esbjorn, 2006; Frame, 2006; McConnell, 2010). It is, thus, difficult to imagine that a setting would be therapeutic or promote learning when the horse is intentionally distressed. When the practitioner is knowledgeable about how horses learn she can better assess the horse throughout the session and ensure that the activities taking place are not contradicting what the horse has already learned.

Pain, fear and high-levels of stress will increase the risk for behaviors that might result in the client, horse, or facilitator becoming injured. Risk-management is central to human-equine interactions and is the practitioner's responsibility (Lombardi, 2014).

Some equine-assisted approaches are based on teaching horsemanship methods to clients, wherein the main challenge is to determine whether the method is in concert with the principles of how horses learn, while others do not explicitly involve teaching of horsemanship. An example of the latter is a scenario where a client is asked to problem-solve around haltering and leading the horse. As the purpose of the session is for the client to practice problem-solving skills, receiving detailed instructions on how to handle the horse ahead of time is not aligned with client goals. However, it can be considered unethical to not help the client understand basic ways that the horse learns and communicates, so that the client, for instance, does not maintain constant, unavoidable pressure (to which the horse has not been habituated or should not be habituated). An example of this would be maintaining constant pressure on the horse's head through the lead rope even though the horse is walking alongside with the client.

Through supporting correct application of learning theory in client activities, practitioners can better promote the wellbeing of all in sessions.

Maintaining Mental and Physical Health in the Therapy Horse

There are many environmental and equine management strategies that can help enhance welfare for the therapy horse, a presentation that is outside of the scope of this article. However, one strategy that is affordable and easy to implement is to provide clear and consistent handling, ensuring that interactions are predictable and controllable to the horse, which reduces stress (Weiss, 1971).

It is important that the therapy horse receives consistent training and handling outside the client session to reinforce desired behaviors and to prevent undesirable behaviors, such as aggression toward humans (McGreevy & McLean, 2010). Through use of learning theory-based training strategies this can be done effectively and ethically. Understanding the principles of how horses learn can also help the EAMH/EAL practitioner assess the methods of other trainers who are working with the therapy horse.

Establishing a Work Role for the Therapy Horse

EAMH and EAL practitioners need to have extensive knowledge in their human field paired with an ability to promote the wellbeing of their clients, therapy horses, and themselves in a very complex environment. Client assessment skills are obtained through the practitioner's "human" degrees and certifications, which are delivered through accredited institutions and as part of continuing education requirements. Selection and assessment of horses for non-traditional equine services such as EAMH and EAL can seem elusive, or, hit-or-miss, without understanding learning theory (along with equine behavior, cognitive capacities and expression of emotional states). Through practical understanding and application of this knowledge, practitioners can assess therapy horses, initially for suitability, and continuously while providing services.

Temperament, defined as early-appearing variation in emotional reactivity in humans (Rettew & McKee, 2005), or as an individual's basic tendencies toward changes and challenges in its environment (Mason, 1984), should be given some consideration when selecting a therapy horse, although we agree with McGreevy (2012) that the overall learning potential of horses is often overlooked.

Establishing a well-defined work role for the therapy horse, that is, reviewing and recording what the horse will be doing and need to know in order to do it, can help the practitioner in determining initial suitability and will guide training to ensure that the horse is prepared for work prior to client contact. Since the therapy horse serves different functions depending on client goals, practitioner methods, and activities used in the session, a standardized work role for the therapy horse in EAMH and EAL is difficult to outline. For instance, the horse can be observed at a distance, or she or he can carry a client who is unaccustomed to riding. The horse may need to be desensitized to certain tools or settings unique to a particular facility or should remain responsive to tension in or body position of the client. This variation in the content of services makes the creation of a customized work role and accompanying training schedule even more important.

CONCLUSION

Understanding learning theory has, as we have presented, a crucial place in effective and ethical EAMH/EAL practice. Our purpose has been to introduce learning theory and its usefulness to equine-assisted practitioners, as it is central to effective risk management, assessment, selection, training, handling, relationship-building, and the quality of client work. Whether the equine-assisted practitioner sees herself as a horse trainer or not, she, in fact, is. Her ability to provide clear, predictable and controllable interactions with the therapy horse, underpinned by learning theory, will greatly affect the welfare of her therapy partner, and ultimately the quality of her client work as the two cannot be separated.

References

Ekholm Fry, N. (2013). Equine-assisted therapy: An overview. In M. Grassberger, R. A. Sherman, O. S. Gileva, K. M. H. Christopher, K. Y. Mumcuoglu (Eds.), Biotherapy: History, principles and practice: A practical guide to the diagnosis and treatment of disease using living organisms (pp. 255-284). New York, NY: Springer http://dx.doi.org/10.1007/978-94-007-6585-6_10

Ekholm Fry, N. (2014, October). How horse culture affects sustainability and risk management. Presented at the annual meeting of PATH International, San Diego, CA.

Esbjorn, R. J. (2006). When horses heal: A qualitative inquiry into equine facilitated psychotherapy. Unpublished Doctoral Dissertation. Institute of Transpersonal Psychology.

Frame, D. L. (2006). Practices of therapists using equine facilitated/assisted psychotherapy in the treatment of adolescents diagnosed with depression: A qualitative study. Unpublished Doctoral Dissertation. New York University School of Social Work.

Goodwin, D., McGreevy, P. D., Waran, N., & McLean, A. (2009). How equitation science can elucidate and refine horsemanship techniques. The Veterinary Journal, 181, 5-11.

Hahn, C. (2004). Behavior and the brain. In P. McGreevy (Ed.). Equine behavior: A guide for veterinarians and equine scientists (pp. 55-85). London: Saunders.

Hanggi, E. B. (2005). The thinking horse: cognition and perception reviewed: Proceedings from the American Association of Equine Practitioners 51st Annual Convention (pp. 246-255). Seattle, WA: American Association of Equine Practitioners.

Innes, L., & McBride, S. (2008). Negative versus positive reinforcement: An evaluation of training strategies for rehabilitated horses. Applied Animal Behaviour Science, 112, 357-368.

Lombardi, S. (2014). Horses, risk, and complexity: Risk management in therapeutic human-equine interaction services. Unpublished Masters thesis. Prescott College.

Mason, W.A. (1984). Animal learning: experience, life modes and cognitive style. Verhandlungen der Deutschen Zoologischen Gesellschaft, 77, 45–56.

McCall, C., & Burgin, S. (2002). Equine utilization of secondary reinforcers during response extinction and acquisition. Applied Animal Behaviour Science, 78, 253-262.

McConnell, P. J. (2010). National Survey on Equine Assisted Therapy: An Exploratory Study of Current Practitioners and Programs. Unpublished Doctoral Dissertation. Walden University.

McGreevy, P. (2007). The advent of equitation science. The Veterinary Journal, 174, 492-500.

McGreevy, P. (2012). Equine behaviour: A guide for veterinarians and equine scientists, (2nd ed.). London: Saunders.

McGreevy, P. & McLean, A. (2007). Roles of learning theory and ethology in equitation. Journal of Veterinary Behavior: Clinical Applications and Research, 2(4), 108-118.

McGreevy, P. & McLean, A. (2009). Punishment in horse-training and the concept of ethical equitation. Journal of Veterinary Behavior: Clinical Applications and Research, 4(5), 193-197.

McGreevy, P., & McLean, A. (2010). Equitation science. West Sussex, UK: John Wiley & Sons Ltd.

McLean A. (2012, September). Habituation: Six methods to harness this natural learning process. Horses and People Magazine.

McLean, A. (2004). Short-term spatial memory in the domestic horse. Applied Animal Behaviour Science, 85, 93–105.

Miller, R.M. (1996). The behavior of the horse. Journal of Equine Veterinary Science, 16(2), 58-59.

Murphy, J. & Arkins, S. (2007). Equine learning behaviour. Behavioural Processes, 76(1), 1-13.

Pavlov, I. P. (1927). Conditioned reflexes: An Iinvestigation of the physiological activity of the cerebral cortex. London: Oxford University Press.

Rettew, D.C., & McKee, L. (2005). Temperament and its role in developmental psychopathology, Harvard Review of Psychiatry. 13(1): 14–27. doi: 10.1080/10673220590923146

Sankey, C., Richard-Yris, M., Leroy, H., Henry, S., & Hausberger, M. (2010). Positive interactions lead to lasting positive memories in horses, Equus caballus. Animal Behaviour, 79(4), 869-875.

Weiss, J. M. (1971). Effects of coping behavior in different warning signal conditions on stress pathology in rats. Journal of Comparative and Physiological Psychology, 77, 1-13.

Wathan J, Burrows AM, Waller BM, McComb K (2015) EquiFACS: The Equine Facial Action Coding System. PLoS ONE 10(8): e0131738. doi:10.1371/journal.pone.0131738.

Made in the USA
Monee, IL
05 November 2026

67977842R00044